Of Seas and Ships and Scientists

'The time has come,' the Walrus said,
'To talk of many things:
Of shoes—and ships—and sealing-wax—
Of cabbages—and kings—
And why the sea is boiling hot—
And whether pigs have wings.'

Lewis Carroll
'The Walrus and the Carpenter'
from *Through the Looking-Glass*

Of Seas and Ships and Scientists

The remarkable story of the UK's National Institute of Oceanography 1949 – 1973

EDITED BY

A.S. Laughton, W.J. Gould, M.J. Tucker and H.S.J. Roe

The Lutterworth Press

The Lutterworth Press
P.O. Box 60
Cambridge
CB1 2NT

www.lutterworth.com
publishing@lutterworth.com

ISBN: 978 0 7188 9230 2

British Library Cataloguing in Publication Data
A catalogue record is available from the British Library

First published in 2010

Unless indicated otherwise, all figures used in this book are the personal property of the authors and their former colleagues or are from the archives of the National Oceanography Centre, Southampton.

Contents

Dedicated to the memory of the
NIO's founding Director

George Deacon

Preface and acknowledgments

In the late stages of World War II the British Admiralty brought together a small group of scientists under the leadership of an exceptional man, Dr G.E.R. Deacon. Their task was to learn enough about sea waves to make predictions of conditions for amphibious landings. Later, Deacon's Group W (for waves) amalgamated with the Discovery Investigations, a whale and Southern Ocean research organization, to form the United Kingdom's first national institute of oceanography with a brief to study all aspects of ocean science. Under Deacon's direction in a remarkably short time the National Institute of Oceanography (NIO) became a world-leading establishment that developed and exploited methods to observe the oceans. In the post-war years, NIO's scientists dramatically improved the understanding of wind-generated waves, tides, deep-ocean currents, the movement of sediments on the continental shelf, the deep ocean floor, the evolution of ocean basins through continental movement and the ocean ecosystem.

The advances made in the NIO era underpin our 21st-century global perspective of the oceans as a critical part of the earth system; the interplay of land, sea, atmosphere and socioeconomics that shapes our world in areas as diverse as climate and weather, water, health, energy and geological resources, disasters, ecosystems, food, biodiversity, and security.

The ocean is a hostile environment that even today presents enormous technical and intellectual challenges. NIO's scientists and engineers, though working with limited resources and, by today's standards, primitive equipment addressed and overcame many of those challenges with ingenuity and perseverance.

Dr John Gould, one of a later generation of scientists at the Institute, became convinced that it was important to document this crucial period in the UK's long history of ocean science. The challenge was taken up by John's colleagues who had worked at NIO and the process of preparing this book started with a meeting at the National Oceanography Centre, Southampton

(NOCS) in October 2006. The project has been coordinated by Sir Anthony Laughton (Director of the Institute of Oceanographic Sciences, the successor to NIO, from 1978 to 1988) and guided by a small editorial team.

We, the editors, have been helped by many individuals; notably the authors who have freely given their time, dug into their papers, revived their memories and contributed chapters. Regrettably, valuable contributions by Dr Peter Herring, Brian Hinde and John Moorey had to be omitted because of space limitations, but the spirit of their recollections has been incorporated in various chapters.

Many others have assisted in the book's production. We are especially indebted to our colleagues Profs. Norman Hamilton and Gwyn Griffiths, and to Martin Harman, a non-oceanographic scientist, for reviewing an early manuscript of the book. Dr Tony Rice, a former member of NIO, gave his advice as an early editor before he was seduced by the joys of lecturing on cruise ships. The staff of the National Oceanographic Library at NOCS, in particular Jane Stevenson and Adrian Burkett, have been extremely helpful in providing access to the archives and identifying and digitising images from the photographic archive. Dr Rory Howlett, Media and Communications Officer of NOCS, gave valuable help by reading the manuscript, advising on style, on procedures for publication and on indexing. We thank the Media Relations Group of the Woods Hole Oceanographic Institution (WHOI) for permission to reproduce images held in the WHOI archives. The Challenger Society for Marine Science kindly permitted the use of Jack Darbyshire's contribution in Chapter 4.

Most of the work of integrating the individual contributions and preparing material for submission to the publishers has been done at home by John Gould who thanks his wife, Hilary, for her support throughout the project and especially for her forbearance during the summer of 2009. John also had the inspiration to choose the title for this book from his knowledge of the poem 'The Walrus and the Carpenter' written in 1872, the same year as the start of the global voyage of HMS Challenger. Finally we are grateful to Prof. Ed Hill, the Director of NOCS, for providing meeting rooms and for assisting financially with travel and publication costs.

Our hope is that this book will interest, educate and inspire its readers and most importantly that it will help to make new generations of ocean scientists aware of the roles played by their predecessors in revealing the oceans' secrets.

Anthony Laughton
John Gould
'Tom' Tucker
Howard Roe

September 2009

Introduction

Margaret Deacon

In southwest Surrey, thirty-five miles from London and twenty-five from the south coast, the Greensand hills rise up to give a view from their modest crests to the Hog's Back in the north and southwards over the seemingly tree-filled depths of the Weald to the more distant chalk hills of the South Downs. Their leafy lanes even today preserve a feeling of remoteness and it is now all but forgotten, except by the survivors, families and friends of those who worked there, that for over 40 years this quintessentially rural English location was the home of a major British scientific institution. Today it appears a somewhat unlikely spot for a research laboratory; even more so for a National Institute of Oceanography (NIO).

Though in continental terms a mere stone's throw from the sea, by British reckoning the 30-odd miles separating the scientists from the element they were studying was indeed a considerable distance. But the Institute, in its choice of location as well as other aspects, arose from initiatives begun during the latter phases of the Second World War, and in its final form it represented a compromise between the aspirations of the different groups of scientists and administrators involved in its creation, and practical considerations. It was largely the latter that led to the choice of an ex-Admiralty building at Witley as the home of NIO. The arrangement worked sufficiently well for research to continue on the site from 1953 until 1995 when the laboratory, by now expanded and renamed the Institute of Oceanographic Sciences Deacon Laboratory, moved to a new purpose-built waterside facility, the Southampton Oceanography Centre (now the National Oceanography Centre, Southampton).

This book tells the story of NIO from its origins until the considerable expansion and first change of name took place in 1973. It begins by showing why there was perceived to be an urgent need in the United Kingdom for an institution of this kind, so much so that proposals were put forward in the early 1940s, while the nation was still at war. To expedite matters, as negotiations between government bodies dragged on, the Admiralty set up its own oceanographic research group (named Group W for waves) in 1944. Group W had already proved its worth by the time the Institute came into being in 1949. This revival of British marine science, especially physical oceanography which had been almost totally neglected in this country during the first half of the century, quickly led to the Institute assuming an active and prominent role on the world stage. It was a time when oceanography was expanding globally as its significance in both war and peace became more widely recognised, while at the same time new ideas and new technologies were opening up exciting opportunities in the exploration and understanding of the oceans. International co-operation in the planning and carrying out of new projects was also becoming increasingly significant and the Institute became the focus for British participation in many such programmes. The contributions of individual scientists, and that of their leader for much of this time, George Deacon, were recognised at home by a cluster of elections to Fellowship of the Royal Society, the UK's highest scientific honour, and by equivalent awards overseas.

Though it throws light on the complex web of political and scientific influences that shaped the development of oceanography during the second half of the 20th century, the book does not attempt to provide a comprehensive account of wider issues. Its purpose is to give a first-hand account of the laboratory itself, and of the individuals who worked there and their scientific contributions, while this is still possible. It has been written almost entirely by surviving former scientific personnel of Group W and NIO who were therefore directly involved in the events related here. Until now NIO's story has been accessible only through the medium of official reports and scientific papers, material that may faithfully record a sequence of events but does not always explain it.

This is a story that deserves to be better known, of a young institution quickly making its mark on a rapidly developing area of scientific enquiry, and helping to lay the foundation for important work still continuing today. Some introductory chapters are also included to explain why, given that British scientists of previous generations had been much involved with the nascent science of the sea, there was not already an institution of this kind here, similar to those already existing in some other countries. These also show how the idea of an institute of physical oceanography, as originally projected, came to be modified during the discussions leading to its formation, so that the National Institute of Oceanography that came into being on 1 April 1949 was a more inclusive organisation in which all major branches of oceanography were represented. As such, it inherited both ships and personnel from the Discovery Committee which had been responsible for a major government-sponsored programme of scientific investigation of the Southern Ocean and whales and whaling in the 1920s and 1930s.

The Historical Context

Britain has a long and illustrious history of scientific ocean exploration, most notably marked by HMS Challenger's global circumnavigation in the 1870s: and yet in the early part of the 20th-century Britain lagged behind the US and many European countries in this field of science. It was the growing realisation of this unfortunate state of affairs, and its consequences for science and the navy in both peace and war, that led to the foundation of a new oceanographic institute being regarded as an increasingly urgent imperative in many quarters. Supporters of the idea believed that this was the best way of remedying the neglect of marine science in Britain during previous decades. Margaret Deacon, daughter of Sir George Deacon and a scientific historian, describes the background to the formation of NIO. Sir George was its first director after leading Group W at the Admiralty Research Laboratory, set up to study ocean waves and how they might influence amphibious landings. Five of this small and diverse group of then-young scientists give their personal memories of what they did.

1

Marine science in the UK before World War II

Margaret Deacon

In the second half of the 17th century Britain enjoyed a great flowering of talent as part of the movement generally known as the Scientific Revolution. Robert Boyle, Robert Hooke, Sir Isaac Newton and Edmond Halley are well-known names from the circle of natural philosophers who joined together in the early 1660s in the Royal Society of London.[1] Among their numerous achievements in many fields it was often overlooked in subsequent generations that all four, and others besides, had been interested in the scientific study of the sea, a reflection of the growing importance of maritime affairs in national life at the time. Currents and tides, depths and water properties, and the instrumentation required to study them were actively investigated, the latter especially by Robert Hooke. Though the difficulty of scientific investigation at sea was perhaps the most enduring lesson from this early work, Halley charted geomagnetism in the South Atlantic and Newton correctly explained the semi-diurnal tide, so accounting for the springs/neaps cycle, whilst Hooke's designs inspired later apparatus.

With such an illustrious pedigree it seems surprising that during the early 20th century oceanography should have been neglected in Britain, at a time when important developments were occurring elsewhere. During the intervening centuries Britain had a significant share in the development of marine science in the era of scientific voyages of exploration that followed Captain James Cook's accurate charting of Canadian waters and his pioneering circumnavigation of Antarctica. By the early 19th century these voyages were increasingly well equipped to study the sea, as well as other aspects of geophysics such as meteorology and terrestrial magnetism.

At home, the data brought back by these and other travellers were collected and analysed by figures like the British geographer and former East India Company surveyor, James Rennell. He spent over 50 years accumulating information on the nature and distribution of ocean currents. Only part of his work was ever published, a volume of charts of the currents of the Atlantic Ocean and an accompanying memoir that appeared posthumously in 1832. He showed that the pattern of surface currents generally followed that of the prevailing winds and it was the force of these winds on the sea that provided the motive power for the current system. This was in contrast to the views of the German geographer Alexander von Humboldt, who believed that surface currents such as the Gulf Stream were part of a more general system of ocean circulation caused by differences in the density of seawater, a theory widely favoured by European scientists. An element of divergence was already beginning to appear between how such questions were viewed in Britain and on the continent of Europe. British pre-eminence as a naval power and in industrial technology during the mid 19th century would enable her scientists to take the lead in the exploration of the oceans for a while but by the early 1900s the initiative had passed to rapidly developing nations on both sides of the Atlantic.

In the United States Humboldt's ideas attracted more interest, gaining wider currency through the writings of Matthew Fontaine Maury. He had already become famous in maritime and business circles for publishing wind and current charts, based on information drawn from logbooks housed in the United States Navy's Depot of Charts and Instruments, of which he was then superintendent. However perhaps the most significant contribution Maury made to the emergent science of the sea was in helping to resolve the vexed question of how to measure reliably the depth of the ocean, which became crucial in the enterprise of laying telegraph cables on the seabed.

On the back of the growing use of steam power at sea, as well as other facets of high-Victorian technology, new methods of investigating the deep sea now began to be rapidly developed. While civil war in the United States temporarily held back developments there, British scientists and their European counterparts took up the scientific study of the deep sea, as cable laying operations in the Mediterranean Sea and the North Atlantic, and soon further afield, opened new lines of enquiry and paved the way for a new era of oceanic investigation.

Among the many interesting discoveries that followed from these developments, the one that caused most surprise in scientific circles was the evidence for the existence of life in the depths of the sea. The first half of the 19th century had been a golden age of discovery for biologists studying the marine life of coasts and shallow seas and the surface layers of the oceans. However it was almost universally accepted that the immense pressure, let alone the cold, darkness and lack of an obvious food supply, would make life impossible at greater depths. When signs of microscopic organisms were found in deep-sea sediment samples most observers concluded that they had lived in the surface layers and that their remains had sunk to the bottom after death. However a few individuals refused to accept this conclusion, pointing to a number of observations that appeared to contradict it. Their case was strengthened when a length of cable in the Mediterranean was lifted for repair after a few years and sessile organisms were found growing on it, although many zoologists were reluctant to abandon an idea that seemed so self-evidently true.

But the Scottish biologist and academic Charles Wyville Thomson was one man open to change. In 1867 he travelled to Norway to examine specimens dredged from 400 fathoms. Inspired by these, together with William Carpenter, a member of the Council of the Royal Society, he initiated expeditions in a venerable naval ship, HMS *Lightning,* and later in HMS *Porcupine,* around the British Isles, finding at even greater depths a rich and varied fauna influenced by variations in sea temperatures. This led Carpenter to ponder the effect of density differences in the sea and to propose the existence of a vast oceanic circulation of water masses heated in equatorial regions and cooled near the poles; a novel idea to British scientists. He conceived the idea of a voyage of circumnavigation to explore the world's oceans and persuaded the government to finance it.

It was, however, Thomson, not Carpenter, who headed the scientific team on board HMS *Challenger*, sailing in 1872 on a round-the-world voyage of marine exploration, and the scientific emphasis shifted back towards deep-sea biology. After the expedition's return the impressive zoological collections were distributed to internationally respected experts on the various groups and their findings published in the fifty-volume *Challenger Report.* This immense task was not completed until 1895 by which time Thomson was dead and his place as editor had been filled by a *Challenger* colleague, the Canadian-born Sir John Murray, who worked on marine sediments. Neither man had a direct interest in the physical results, and the deep-sea temperatures and other data were only cursorily reported. It was some 20 years before this information was examined more critically by German scientists who included it among data they used to plot the deep circulation of the Atlantic, producing a picture far more sophisticated than the earlier hemispheric model adopted by Carpenter. In its time the *Challenger* Expedition was an outstanding success, both scientifically and in the realm of international public relations, but the focus for new development soon shifted elsewhere.

During the 1870s and '80s, in nations where fisheries were economically important, scientists drew attention to the possible effects of weather and sea conditions on food fishes and their distribution. Swedish work on the herring industry led to the creation of the International Council for the Exploration of the Sea (ICES) in 1902[2] to undertake joint investigations of the physics, chemistry and biology of the North Atlantic. At about the same time observations made on Arctic expeditions and, in particular Nansen's epic voyage in the *Fram* (soon to be supplemented by data collected on Antarctic expeditions in the early 1900s) led to the development of a dynamical approach to the study of ocean circulation.

Meanwhile in Germany, studies of ocean circulation of a more traditional, qualitative, nature were in progress at the Deutsche Seewarte in Hamburg and the Institut für Meereskunde[3] in Berlin, developing new ways of understanding how cold deep water of Antarctic origin found its way into the Atlantic Ocean. The German researches and the results of the *Meteor* Expedition of 1925-28 were incorporated into Albert Defant's 1929 textbook, *Dynamische Ozeanographie*, which introduced the dynamical method to a new generation of oceanographers: one of these was George Deacon, later to become the first director of NIO.

The first institutions for marine research were zoological laboratories, or marine stations, that appeared on both sides of the Atlantic from the mid 19th century onwards. Many began as small seasonal foundations by individuals and scientific societies and typically remained attached to university departments, providing research facilities for professors and their students, and often for visiting biologists. Some, however, such as the Stazione Zoologica at Naples (founded by the German zoologist Anton Dohrn in 1872) and the Marine Biological Laboratory at Woods Hole (1888) became important centres not just for the study of life in the sea but also for many wider aspects of biological research, for which marine organisms proved well suited.

By the early 1900s however it was beginning to be appreciated that to advance the study of oceanography it would be necessary to develop institutions in which the different disciplines were better represented. Some European initiatives have already been mentioned above but there was also growing awareness of this need among scientists in the United States.[4] On the west coast this was addressed by adapting an existing foundation. Following the retirement in 1923 of its founder W.E. Ritter, the Scripps Marine Biological Station at La Jolla, an arm of the University of California, was assisted by the Carnegie Institution to become, in 1925, the Scripps Institution of Oceanography, and in 1936, a leading Norwegian oceanographer of the Bergen School, Harald Sverdrup, was recruited as director. On the east coast, a new foundation, the Woods Hole Oceanographic Institution was established in 1930 adjacent to the Marine Biological Laboratory, through the agency of the National Academy of Sciences. Lack of resources restricted activity prior to the USA's engagement in the Second World War but some important developments were made. Furthermore, *The Oceans*, by Sverdrup, Johnson and Fleming, published in 1942, provided an account in English of the then-state of oceanography and proved influential as a blueprint for the subject's development after the war.

By 1939 oceanographic institutes had also been established in Russia, Norway, Sweden and Japan, but not in the United Kingdom. Britain was active in some areas of marine science, but the nation that sent out the *Challenger* Expedition seemed largely to have lost interest in oceanic research, especially physical oceanography, in the first part of the 20th century. Was this due to lack of interest or to lack of opportunity? It was undoubtedly true in the early 1900s that the general public in Britain were more willing to support polar adventures with a veneer of scientific respectability rather than more prosaic field research programmes. Until well after the Second World War oceanography scarcely impinged on the public consciousness, to an extent that is almost inconceivable today. It may not have helped that those anxious to promote this field of research were mainly geographers and biologists, rather than mathematicians and physicists who would have found the new developments in dynamical oceanography more accessible. As a result only a few specialised groups and individuals preserved the vision of a wider approach to marine research.

The main problem was money. The *Challenger* Expedition had public support at the time it was sent out but this dwindled during its long absence and turned to active hostility, even among some scientists, as the production of the report, largely at public expense, dragged on. Though the UK had scientific patrons, they were not on the scale of the big North American foundations or of individuals such as Alexander Agassiz in the USA or Prince Albert I of Monaco in Europe. John Murray eventually became a rich man through the extraction of phosphates from Christmas Island and left money to support marine research, but his Scottish Marine Station of 1884, the only 19th-century British foundation carrying out broadly based marine research, had long since failed because he could attract neither private nor public funds to keep it going.[5] Similar problems were encountered by other marine stations elsewhere in the UK about this time, and their existence was often precarious.

However a lifeline was occasionally available through government grants to investigate fishery questions. In the late 19th century the prevailing view was that the state should not fund basic scientific research, but that in matters of public interest it could be justified. Problems being experienced by British fishermen, facing declining catches in some species and anxious about possible effects of trawling and pollution, led to grants being made available for studies of fish biology. Some of this was handled by government departments in Scotland, Ireland (pre-independence) and England but private organisations also benefitted. Most successful in this respect was W.A. Herdman, professor of zoology at Liverpool University from 1881 who ran the marine station at Port Erin in the Isle of Man, as well as a laboratory at Liverpool. In 1920 he also established and briefly held a chair of oceanography there, intended for fisheries research.

From the 1920s onwards, a somewhat more liberal approach assisted the struggling independent marine laboratories. Among those that benefitted were the Plymouth Laboratory of the Marine Biological Association of the United Kingdom and the Scottish Marine Biological Association's station at Millport on the Firth of Clyde. Grants from the Development Commission enabled extra staff to be taken on in these two institutions and made possible important interdisciplinary work on productivity in the sea, with chemists assisting the biologists.[6] Between the wars government-run fisheries laboratories in England and Scotland also employed physical oceanographers or hydrographers as they were then termed. The feeling grew however that by making physical oceanography subservient to the biological aims of these bodies, not only was the discipline itself being held back, but also its potential input to other aspects of marine science. By the onset of the Second World War it was generally agreed by those involved in marine research in the United Kingdom that its progress here had been badly distorted by the overemphasis by government agencies on applied fisheries research.

However there were signs that attitudes were changing. This was partly due to the activities of the Challenger Society for the Promotion of the Study of Oceanography, founded in 1903 by a group of British marine scientists headed by G. Herbert Fowler,[7] a zoologist working at University College, London. Though its membership consisted mainly of marine biologists, the society welcomed anyone interested in the science of the sea. They compiled a manual, *Science of the Sea*, which appeared in 1912, to help the private yachting community to make observations. In the short term such hopes were dashed by the outbreak of war two years later, but the book had some influence on thinking about marine science in the post-war period. Most notably it was read by H.G. Maurice, a young civil servant, who, as Assistant Secretary at the Board (later Ministry) of Agriculture and Fisheries from 1912 to 1948, was himself an influential figure in marine science during this period. He paid tribute to the book's role in helping him develop a broader understanding of the aims and needs of marine research.

During the First World War Fowler, who had retired from academe, worked as a volunteer for the Admiralty compiling charts for use by submarines. The Navy was already developing acoustic submarine detection and it was soon evident that the operation of devices was affected, among other things, by variations in the velocity of sound in the water, a function of its salinity and temperature. Fowler represented to his superior, the Hydrographer of the Navy, that naval vessels after the war should make routine observations of this kind and an oceanographic branch might be set up in the Hydrographic Department to collect and process the data. The Hydrographer of the time was not enthusiastic about this idea, but others took Fowler's suggestions more seriously, even including adapting a submarine to make oceanographic observations. This proposal, ahead of its time, ended in an onboard mutiny and was abandoned, but again there were positive results. To develop the necessary apparatus for the trials Fowler had recruited D.J. Matthews who had worked at Plymouth before the war. This time, when the project ended prematurely, the Admiralty's Director of Research, F.E. Smith, kept Matthews on in the Hydrographic Department, and he continued to work there until retiring in 1936. Matthews is principally remembered for his tables of the speed of sound in seawater as a function of temperature, salinity, pressure and geographical location.

This episode reflects a growing awareness among Admiralty personnel engaged in scientific and weapons research and development, that they needed to pay more attention to oceanography. Submarines were employed on occasion in the 1920s and 30s to collect salinity and temperature data in the Mediterranean to investigate the effect of local sea conditions on the performance of ASDIC (sonar) apparatus. From 1933 onwards the emphasis on naval research in underwater acoustics shifted away from the development of detection apparatus to 'sea research', examination of the factors affecting its performance.[8] This work was based at HMS *Osprey*, an Admiralty shore-based establishment at Portland, and was, in the mid-1930s, in advance of parallel researches being carried out in the USA.[9] However it was at Woods Hole in 1937 that oceanographers solved a problem encountered by the US Navy of deteriorating sonar signals on fine days. They found it was produced by the refraction of sound in layers of water near the sea surface warmed by the sun, and shortly afterwards perfected a new kind of apparatus capable for the first time of measuring continuous temperature profiles in the surface waters, the bathythermograph.

In other areas of national activity, as in naval science, situations were also arising which drew the nation more actively into marine exploration, in ways that were not always foreseen in advance. By the late 1930s marine scientists from both sides of the ocean were drawing up proposals for joint explorations in the Atlantic, to include both hydrographic and biological work. The British fisheries hydrographers were involved in these discussions under the umbrella of ICES and a naval vessel was to be made available for observations in the mid-Atlantic. Meanwhile the International Association for Physical Oceanography was also planning for an international survey of the Gulf Stream and the Royal Society agreed that British workers should co-operate with American scientists in this investigation.

One can only speculate what effect these international engagements might have had in promoting oceanography at home. In the event, the outbreak of war in 1939 put paid to such proposals, at least for the time being. However it did not prevent discussions about the best way forward after the war was over, in which the need for an oceanographic research institute increasingly featured. Also, although it was agreed that the nation's record in the study of oceanography during the first half of the 20th century had been generally poor, there had been two notable exceptions, both of which had gained international standing through their work, and both of these had a significant role in events leading up to the foundation of the National Institute of Oceanography, and in its development.

One was the transformation in the mid-1930s of Herdman's Liverpool Chair of Oceanography, from fisheries biology to physical oceanography. Its new holder, Joseph Proudman, had become interested in the study of oceanic tides while Professor of Applied Mathematics at the university and established the Tidal Institute in 1919. This later moved across the Mersey to the Bidston Observatory under his deputy, Arthur Doodson, and gained an international reputation for its work.

The other outstanding development at this time was a programme of research in Antarctic seas, the Discovery Investigations, carried out by British scientists in the 1920s and '30s, under the auspices of the Discovery Committee in relation to the whaling industry. Within a few years of the discovery of the sub-Antarctic islands early in the 19th century, the land-based biological resources of the Southern Ocean, particularly the seals, were being heavily overexploited. In contrast, the Southern Ocean whales were not targeted significantly until the beginning of the twentieth century because of the logistical problems involved and the lack of relevant technology. Modern Antarctic whaling began when the veteran Norwegian sea-captain C.A. Larsen established the first land-based whaling station in the Southern Ocean, at Grytviken, South Georgia, in 1904. Others soon followed.

The numbers of whales taken increased rapidly over the next few years so that as early as 1910 concern was being expressed both at the wastefulness of the industry, since large parts of each whale carcass were discarded, and at the potential effect of the fishery on the whale populations. These concerns were partly what we would now call conservationist, but were also based on financial considerations. The Falkland Islands Dependencies had been formally established by the Colonial Office in 1908 with, *inter alia,* responsibility for regulating the whale fishery and to derive an income from it by levying duty on all whale oil passing through its territory, at that time specifically South Georgia. Clearly, the long-term security of such an income would depend upon limiting catches to a sustainable level. The problem was that no-one knew what a sustainable level was.

The deliberations of the first Interdepartmental Committee set up to consider the concerns were interrupted by the First World War, but a second committee to consider 'Research and Development in the Dependencies of the Falkland Islands' was convened in April 1918 which recommended a tripling of the duty on whale oil from the Dependencies and the establishment of a programme of research in the Southern Ocean from dedicated research vessels. After some prevarication, a third committee recommended the purchase of Scott's *Discovery* for this purpose and the construction of a new vessel, the RRS *William Scoresby.* These proposals were accepted and the Colonial Office set up the 'Whaling Research Executive Committee' to oversee the work. It met first in April 1923, but after the purchase and refitting of the *Discovery* it became known as the Discovery Committee, established 'for the purpose of conducting research into the economic resources of the Antarctic and with the particular object of providing a scientific foundation for the whaling industry'. The programme of research became the Discovery Investigations.[9]

It was not at all unusual for the Colonial Office to support fisheries research in British overseas possessions, and many young marine biologists found employment on such projects. What was unusual in this case was that what was supposed to have been a single expedition developed into a research programme of many years duration.

Stanley Wells Kemp, in the early 1900s, had been a member of E.W.L. Holt's team of gifted young researchers in the fisheries section of the Department of Agriculture and Technical Instruction for Ireland. He was both an excellent administrator and an inspiring leader. When it became clear that the work of the *Discovery* expedition of 1925-27 was only a beginning, he planned further work and persuaded the Committee that a new ship was essential. With remarkable speed, a purpose-built steel ship with oil-fired steam engines, the Royal Research Ship *Discovery II*, was launched in time to sail for the Antarctic before the end of 1929. At 234 feet she was longer, faster and roomier than her predecessor, and equipped with moderate ice-protection. Between 1929 and 1939 RRS *Discovery II* carried out five Antarctic commissions, with a sixth after the war when she was with NIO. The *William Scoresby* carried out seven pre-war commissions, and an eighth post-war, not all of her work being done in the far south.[10] Much was learned, not only about the biology and ecology of whales, but also about the Southern Ocean that sustained this remarkable ecosystem. This was due in large part to the Committee's choice of scientific director. The work also introduced a new generation of British scientists to oceanographic research, including a young chemist, George Deacon, who would play a major part in the development of NIO.

2

Steps toward the founding of NIO

Margaret Deacon

Naval personnel in what would later become the Royal Naval Scientific Service were well aware of the value of oceanographic work to their concerns. However, the events leading to the founding of NIO were actually set in train by the Hydrographer of the Navy, Sir John Edgell. In the summer of 1943 he was asked to attend a meeting of the Scientific Advisory Committee (SAC) to the War Cabinet. The reason was that a leading Swedish oceanographer, Hans Pettersson,[1] had recently approached the British government via an intermediary to ask if it would be interested in a joint expedition to study the Atlantic seabed after the war. At the meeting on 14 September, as well as giving the Admiralty response, Edgell '*took the opportunity of saying that, on general grounds, he would welcome the establishment of an oceanographic institute in this country*'.[2] He further '*undertook to submit to the Committee a Memorandum outlining the earlier history of the proposals for mapping the Atlantic seabed, and giving his views on the practicability of the particular proposal before the Committee, and also on the more general question of recommending that steps should be taken for the setting up of an oceanographic institute*'.

Edgell[3] was no stranger to oceanography. Born in 1880, he joined the Navy as a boy and after entering the surveying service in 1902, rose to be Hydrographer in 1932. In this role Edgell was in close contact with marine research in the UK, both in the Navy and other organisations. He was a member of the Discovery Committee and numerous other relevant committees, including the Oceanography Subcommittee of the British National Committee for Geodesy and Geophysics. His department had close links with the Tidal Institute at Liverpool and he was on friendly terms with Proudman, its founder and director. He also knew the hydrological work being done by the fisheries departments and the people carrying it out, and had asked for one of them, J.N. Carruthers of the Fisheries Laboratory at Lowestoft, to be seconded to his department if war broke out. In 1938 he had been instrumental in proposals to ICES for investigations of the Atlantic seabed, and secured agreement that a naval survey vessel would be available for work north of the Azores for limited periods in 1940 and 1941.

Vice Admiral Sir John Edgell, KBE, CB, FRS. Hydrographer of the Navy. He was a major influence in the steps to establish the NIO. (By courtesy of UK Hydrographic Office Archive.)

It seems unlikely that Edgell's proposal for an institute was long premeditated, and it appears to have been put forward as a personal suggestion, rather than as one having official Admiralty backing. In preparation for his appearance before the Committee, he had asked Joseph Proudman, as one of the most senior figures in UK marine science, for his opinion of Pettersson's proposal. Proudman[4] advised against participation in the Swedish expedition and the Swedish Deep-Sea (*Albatross*) Expedition of 1947 took place without British involvement. He felt that permanent institutions, such as Woods Hole, were a better use of the scarce resources currently available, but recommended that Edgell seek the opinion of marine scientists from the younger generation, George Deacon[5] and Edward Bullard.[6]

This letter from Proudman seems the most likely origin of Edgell's proposal. If the idea was new it would have immediately appealed to him, for one consequence of the war was that the Hydrographic Department had been deluged with requests for oceanographic information, from civilian bodies as well as from the armed services, far more than Carruthers, acting as his assistant, could cope with almost single-handed.

It was not only the Royal Navy that was concerned about this state of affairs. By the mid-1940s a general movement was afoot to think ahead to what scientific priorities should be after the war ended. There was concern that more attention should be paid to fundamental research that had, of necessity, been neglected during the emergency. A proposal for a national institute of physical oceanography had recently been put forward by the Scottish fisheries hydrographer, John B. Tait, in a 'Memorandum on the significance of scientific research on planning for the post-war reconstruction of the fishing industry'.[7] In this he argued that this step was necessary to rescue physical oceanography from its subordinate role in British fisheries science, which he felt was holding back its progress. This memorandum had clearly made an impression on government and later that year E.H.E. Havelock of the Development Commission (one of the government bodies financing marine research in Britain in the first half of the 20th century) asked the Cambridge mathematician Sir Geoffrey Taylor for his opinion of the idea. Taylor's reply[8] showed that there had also been considerable discussion of the future of oceanography among Cambridge geophysicists who were keen to see the seismic techniques introduced from the USA by Edward Bullard before the war applied to the study of the Mid-Atlantic Ridge. They too felt that an oceanographic centre ought to be established, either as part of a national institution for geophysics or independently.

The SAC invited Edgell to submit a memorandum of his views, which he duly did,[9] and in January 1944 it discussed the matter and decided that both proposals, for an expedition and an institute, should be referred to the National Committee for Geodesy and Geophysics. This committee, organised by the Royal Society as part of the national contribution to the work of the International Union of Geodesy and Geophysics, had an oceanography subcommittee of which Edgell, Proudman, Matthews, Tait, Carruthers and Deacon were already members.

Several documents had been prepared before the meeting, chaired by Edgell on 1 March 1944. Back in December Proudman had drawn up a plan of his suggestions for the work of an institute.[10] He believed that it should cover physical oceanography only, as marine biology was so extensively provided for elsewhere, and that it should be located at Liverpool. He continued to lobby for this solution during the protracted negotiations that followed. Carruthers had more modest expectations of what might be achieved, probably the result of long years of hopes deferred. He thought that the institute should act in the first place as a data centre and that they would never get the funds to run and fully use an ocean-going research vessel, but he did think that there should be a biological presence.[11]

George Deacon presented a paper, 'Oceanographical Research',[12] that was less a blueprint for an institute than for the oceanographic research it should foster. He emphasised the importance of co-operation between the various branches of marine science, arguing that such an institute ought to cover physical oceanography, marine biology, marine sediment studies and chemistry. He concluded:-

> *The most rapid advance in any of the branches of oceanography will be made by orderly, intensive and concerted attack on one or other aspect. Expeditions are needed for filling gaps and exploring areas from which only scattered data are available, but the need is even greater for systematic work by well-equipped stations and research vessels that will represent all the marine sciences so that findings in different fields can be correlated.*

The fact that much of the support for oceanographical work has for a long time been secured because of its application to fishery problems has tended to allow the relegation of physical work to a secondary position. This is considered by most physical oceanographers to have retarded the advance not only of the physical problems, but also of the biological problems that it was hoped to further. This neglect is not so obvious during the past 10 years, but it may still be made good; in the ideal fishery investigation the protracted enquiry into natural history and physiology must include all the physical and chemical work necessary to follow the whole life history of any species. It may be more difficult to obtain financial support for such an enterprise because it is not possible to say in advance what results will be obtained, and many problems must be attacked which seem remote from practical application or economic bearing. A long period is needed for the work to reach a productive state.

Edgell was in agreement about the inclusion of marine biology but he also thought that some of the others were not being bold enough in their expectations. He had earlier written to Carruthers:-

The more I go into this idea of an Oceanographical Institute, the more interesting it becomes, and I am beginning to have quite decided views on its make-up. I am inclined to think that you and Proudman, Tait and perhaps G.I. Taylor also, are thinking too much in terms of the British Isles and surrounding waters; my own ideas are much more ambitious and where you speak of spending £5,000 to £10,000 a year, I am much more inclined to think of £30,000, for I believe that unless we go for a maximum scheme we shall defeat our own object.[13]

He put this view forward even more forcefully at the meeting:-

My own view is that unless the Oceanographical Institute is run on generous lines, and provided with ample funds, it will fail to achieve its object and I would rather try to establish a major organisation than one which has to live on starvation rations. I know that this large view is not shared by all members of the Sub-Committee, also that it will be extremely difficult to get the necessary money, none-the-less I should hope for the setting up of an Institute with a suitable vessel attached at an annual cost of £50,000.[14]

On the Swedish proposal the general feeling of the meeting was that, as there was no realistic possibility of such a project getting off the ground until after the cessation of hostilities worldwide, they should at present concentrate on plans for an institute. All present then declared themselves in favour of the establishment of an institute and the meeting proceeded to discuss the proposal in greater detail. It was then agreed in principle that there should be a junior biologist on the staff to liaise with other institutions. At a follow-up meeting in May[15] this position was upgraded, and on Bullard's recommendation it was decided to include a geophysicist rather than a geologist. There was also to be a meteorologist but the senior posts would be in physics and chemistry. The committee's report was subsequently drawn up by Edgell and submitted via the Royal Society to the SAC later that year.

Fortunately for the future of the science the Admiralty was not prepared to wait. In June 1944 an Oceanographic Research Group was established at the Admiralty Research Laboratory in Teddington with George Deacon at its head. Group W (for waves), as it was generally referred to, was set up to improve understanding of the physics of waves at sea. This was a subject that had previously proved intractable. The problem of wave forecasting for amphibious landings had been tackled with some success here by the Naval Weather Service,[16] and important contributions in this area were also being made by oceanographers Harald Sverdrup and Walter Munk in the USA.[17] Both nations co-operated in the Swell Forecast Section in the run-up to D-Day but this organisation was subsequently transferred to the Far East. Group W's role was to investigate the basic processes involved, on behalf of the Navy. However its future success in establishing the science of sea waves and how to forecast them would have important applications in peace as well as war.

Meanwhile over the next few months the SAC discussed the Edgell Report with scientific representatives and senior civil servants from interested departments. They looked at various ways it might be financed and how it should be governed, and decided that the £50,000 a year required was a legitimate charge on public funds and that it should be located at Liverpool but have a status independent of the university. The committee then unanimously agreed to forward its recommendations to the government.

Until this time there had been no suggestion that plans for the new institute should in any way be linked with the fortunes of Discovery Investigations, but this possibility was raised in the summer of 1945 by A.V. Hill who 'understood that the Colonial Office were anxious to be relieved of their responsibility for the Discovery Committee'. To avoid multiplying administrative bodies in this field, he suggested that the governing body of the institute should look after both organisations. Edgell raised no objection, apart from stipulating that the institute's research vessel should not be used for polar work.

During the latter part of 1943 the Discovery Committee had also been looking to the future. Neil Mackintosh and members of its scientific subcommittee had continued to meet during the war and put forward proposals for the resumption of work that were endorsed by a meeting of the full Committee on 6 June 1944. While the Committee recognised that its initial function, the scientific study of whaling, was no longer of such significance to the Colonial Office, it felt that while doing such work Discovery Investigations had acquired a more general knowledge of the Antarctic regions that would become valuable after the war, when improved communications would render the area more accessible. It felt that this justified asking for new funding and a wider remit but the Colonial Office was unwilling to agree. In 1945 it attempted to transfer Discovery Investigations to the Department of Scientific and Industrial Research but this proposal was robustly resisted by that organisation as unsuitable. It was at this point that the SAC recommendations and the Edgell Report arrived on various Whitehall desks, and others as well as Hill saw the possibility of a tidy solution to the problem.[18] However nearly five years were to pass before the National Institute of Oceanography was actually established. In 1945 the war ended and Edgell, whose tenure as Hydrographer had been extended long beyond the norm, retired. The backlash from the concentrated efforts of the war years caused a general lessening of confidence and enthusiasm, and cut-backs in the research budget. It was amazing that the whole project did not get lost in the labyrinthine discussions that followed both within and between government departments.

It was July 1946 before the SAC received the government response to their proposal. This came partly in the form of a Treasury memorandum stating that the institute's relationship to the Discovery Committee should be settled. The Committee had recently put forward proposals for research activity over the next five years, at an estimated annual cost of £50,000, and the Treasury was not prepared to fund both bodies. As the Discovery Committee was an established organisation with accumulated experience and goodwill there was a case that it should continue rather than be subordinated to a 'new unknown and untried body'. The Treasury therefore proposed that the Discovery Committee and the institute should be merged into a single body in order to economise on costs, and to provide a balance between the interests of physical and biological oceanography.[19]

The SAC eventually agreed with the Treasury that there would be great advantage in placing general responsibility for ocean research on the Discovery Committee. It appeared that much research remained to be done in the Southern Hemisphere as well as urgent need for oceanographic research elsewhere in the world but that Discovery Committee vessels could do this. The Committee therefore recommended that the reconstituted Discovery Committee should be transferred from the Colonial Office to the Admiralty and have responsibility for all bodies interested in oceanography. The Discovery Committee should remain in London

Dr N.A. Mackintosh, who succeeded Kemp as head of the Discovery Investigations. He was Deputy Director at the founding of NIO.

but this did not invalidate the idea of the institute being in Liverpool. The interest in oceanography of the dominions, India and the colonies should also be borne in mind.

In March 1947 the Treasury recommendations for the foundation of the institute were accepted, with minor changes, by the Advisory Council on Scientific Policy, the post-war successor to the SAC, and it might have been expected that the way ahead was now clear. However this proved not to be the case and in August 1948 its chairman, Sir Henry Tizard, wrote to the Admiralty asking why no further progress had been made in setting up the Institute.[20] He was told that this was because the Admiralty could not meet the full cost and that expected assistance from other departments had not been forthcoming. Perhaps because of his family link with oceanography, his father, T.H. Tizard, having sailed in the *Challenger*, Tizard was influential in trying to get things moving.

In fact a great deal of heart searching had gone on among the various departments of the Admiralty where there were sometimes conflicting views on what form the institute should take. There was the belief, sincerely held by many, that it would be improper for the Royal Navy to take over an organisation such as the Discovery Committee whose work had little to do with defence. Yet they were being told by the Treasury that they must finance the whole package out of the naval vote, without extra funds, or face the prospect of losing the institute. Much of the responsibility for the stalemate lay with Sir Alan Barlow, the Second Secretary at the Treasury. Though not unsympathetic to science (he was married to a granddaughter of Charles Darwin) Barlow had traditionalist views about spending public money on it.[21] Fortunately other counsels prevailed and by the end of the year the difficulties had been largely resolved, the final details being approved by the Treasury in February 1949.

The Discovery Committee was to be wound up and a National Oceanographic Council created by Royal Charter 'with the object of advancing the science of oceanography in all its aspects'. This body was to work through an executive committee, very much as originally recommended by the Edgell committee. The institute, which was to cover both physical oceanography and marine biology, would receive financial support from the Development Commission and the Colonial Office, and from Commonwealth governments, but it would principally be funded by the Admiralty, £50,000 being set aside in the first year. The Admiralty also purchased the *Discovery II* and *William Scoresby* from the Government of the Falkland Islands and presented them to the institute. The Discovery Committee was disbanded in March and the National Institute of Oceanography came into being on 1 April 1949.[22] George Deacon, the preferred candidate of the Royal Naval Scientific Service, was appointed Director some weeks later.

Meanwhile, the question of where the institute should be located was still unsettled. For the time being its component parts remained scattered; the Oceanographical Group of the Royal Naval Scientific Service (Group W) continued at the Admiralty Research Laboratory, Teddington; Discovery Investigations scientists at the Natural History Museum; and the Oceanographic Branch of the Hydrographic Department at Cricklewood in North London. Since Proudman's first suggestion in 1943 it had been intended to establish the institute at Liverpool but Neil Mackintosh vehemently resisted any plan to move Discovery Investigations out of London. The real necessity however was for an existing building to be found that could accommodate all sections of NIO on one site, as a new building would be too expensive. Options were considered from Scotland to the South Coast but the consensus increasingly was that it should be in the London area where so many of the staff were already living.

Early in 1950 the committee strongly recommended the purchase of Ridgemead, a pre-war Lutyens-designed mansion at Englefield Green, but was unable to proceed before the Council was in place and the site was sold to another purchaser. However by August the possibility had arisen that they might later be able to acquire cheaply a large wartime Admiralty building at Witley in Surrey.[23]

On 9 October 1950 the Royal Charter incorporating the National Oceanographic Council was approved by Order in Council, and the Council, which included many names previously mentioned in these pages, first met in February 1951. One of its first actions was to constitute the Executive Committee, which till then had been provisional. The first chairman, Sir Frederick Brundrett, who as head of RNSS had done so much to bring the plans for NIO to fruition, had been transferred to the Ministry of Defence and was replaced by his successor, W.R.J. Cook. Among the other new appointments was Vice-Admiral Sir John Edgell, a fitting recognition of the part he too had played in the institute's foundation.

The Council authorised the Committee to acquire permanent premises for the institute and by the summer of that year plans for the move to Surrey were already being drawn up. NIO was to lease the building, originally erected by the Admiralty in 1943 as an extension of the Signal and Radar

The NIO building in the mid 1950s. The big black roller doors were later removed.

Research Establishment, then located at Haslemere. It stood in the grounds of King Edward's School, Witley, in a semi-rural situation (sometimes referred to as Wormley, the name of the local telephone exchange) but only a short walk from the local Witley railway station. The building itself made an ideal home for the young institute; it was plainly built but strong and serviceable. The situation was to some extent a compromise. The advantages of easy rail access to London and Portsmouth, for both staff and visitors, promoting links with scientific colleagues and making it easy to attend society meetings, were felt to outweigh the fact that it was 25 miles from the sea. The argument went that a central position with a choice of ports had much to recommend it in a small country like the UK. A seashore location was not necessarily an advantage if one was dealing with deep-water science. The work of readying the new building occupied a further two years but in the spring of 1953 the move at last took place and the staff settled in to continue the work which had already been in progress for several years and which is described in the following chapters.

Note to the reader

Much of the information contained in this chapter is based on unpublished material in The National Archives at Kew (TNA), including Cabinet (CAB), Admiralty (ADM) and Colonial Office (CO) papers. Other important sources are in the Hydrographic Office (HO) at Taunton (Ministry of Defence), and the National Oceanographic Library (NOL) at the National Oceanography Centre, Southampton, which holds the papers of George Deacon (GERD)[24] and other NIO scientists.

3

The founding director, Sir George Deacon

Anthony Laughton and Margaret Deacon

George Edward Raven Deacon (later Sir George Deacon) was the founding Director of NIO, an inspiration to all the oceanographers who worked under him, and a leading light in the development of oceanography. He left unpublished notes[1] for the Royal Society about his life, which have been used in this chapter.

He was born in Leicester in the English Midlands in March 1906, the second child and only son of George Raven Deacon and his wife Emma, née Drinkwater, his sister Grace having been born two years previously. His parents were strict Baptists, and he remained throughout his life much influenced by the values inculcated during childhood, both moral and practical. As they were not at all well off, holidays were occasional day trips to the seaside. It was on one of these that young Ted, as he was known at home, was excited by the prospect of a boat trip across the Humber, and greatly disappointed when the trip was called off owing to his sister's nervousness. He later joked that it was this disappointment that made him want to go to sea.

Despite limited means, Deacon's parents were anxious for their children to have the opportunity of higher education and Grace became the first graduate of Leicester University to obtain a first-class honours degree in mathematics. Maths was also Deacon's first preference at the City Boys School. He always spoke with great appreciation of the teaching he had there, and the mathematics master, Bert Carpenter, in particular became a lifelong friend. He followed Carpenter's footsteps to King's College, London, where he took a first-class degree in chemistry (because it gave wider employment prospects) in 1926, the year of the general strike.

He returned for a final year in 1927 for the education course, teaching two days a week at Kilburn Grammar School but doing some chemistry, mostly in the evenings. At the end of the year he got his diploma in education and teacher's certificate, and published a paper in the Journal of the Chemical Society on the possibility of double compounds of lead chloride and sodium chloride, and of lead chloride and lithium chloride. His next paper a few years later was on the northward spread of Antarctic water into mid-latitudes of the Atlantic Ocean.

In 1927, Deacon replied to an advertisement by the Discovery Committee for a chemist to work in a small ship in Antarctic waters. He had no reply, "*probably because the old square-rigged Discovery was creeping slowly home from her 1925-1927 voyage*". Instead he went to teach chemistry and mathematics at Rochdale Technical School. A few months later he was offered and accepted the post in the *William Scoresby*, a 123-ft long, 'salt beef and salt pork' ship.

In the *William Scoresby*, and in South Georgia, Deacon worked with zoologist colleagues D. Dilwyn John, F.C. Fraser and J.W.S. Marr and began to appreciate how the distribution of marine organisms, from the great whales to the small creatures on which they fed such as krill, was governed by the physical conditions of the seas they inhabited, and the effect of density differences on water movements. He was glad to have the chance to talk to the young Håkon Mosby when the *Norvegia* called at Grytviken. This meeting, and reading as widely as he could, introduced Deacon to the new ideas on ocean circulation being developed on the European continent. In the literature of the German and other Antarctic expeditions, that first described the circulation of Antarctic waters, and from the results of the recent *Meteor* expedition, he learnt how Atlantic and Antarctic water masses interweave at different levels in a complex gravity-driven pattern determined by their physical characteristics.

He went south again in 1930-31, joining *Discovery II* at Cape Town during her first commission, and after his return completed a monograph on the South Atlantic Ocean,[2] in which he confirmed and extended earlier work, before sailing south again in September 1931. This marked an extension of the Committee's operations as pelagic (i.e. open ocean) whaling was now taking over from land-based stations so that the whalers were no longer tied to island bases and *Discovery II* was visiting new areas. The 1931-33 commission involved often gruelling and sometimes dangerous work in icy seas as the ship completed the first winter circumnavigation of Antarctica.[3] This gave Deacon the opportunity he needed to extend his own researches to new areas of the Southern Ocean. His important monograph of 1937, '*The Hydrology of the Southern Ocean*',[4] was the first work to look at this ocean as a whole, and to show how the circulation patterns known in the Atlantic were also present in other sectors. Perhaps most significantly he drew attention to the wider significance of Antarctic water in the general ocean circulation. Henceforward physical oceanographers would have to think about global circulation, rather than treating the individual oceans as separate entities.[5] The work was ready for the press before Deacon sailed again, this time as chief scientist on *Discovery II*'s fourth commission (1935-37). The scientific programme of the early part of this voyage was disrupted when *Discovery II* was diverted to the Bay of Whales in the Ross Sea to search successfully for the American explorer Lincoln Ellsworth, and his pilot, who were missing after making the first trans-Antarctic flight. Before they returned Stanley Kemp had left to become Director of the Marine Biological Association Laboratory in Plymouth, much to Deacon's regret as he greatly valued his leadership.

As the war approached, Deacon finished a note on carbon dioxide in the Antarctic ocean and joined the Admiralty's anti-submarine research team attached to HMS *Osprey* in Portland. Mainly on HMS *Kingfisher*, he helped in research and development for new methods and equipment for the detection of submarines in training and dummy mines moored in West Bay. He also had to see whether presentation of acoustic echoes on an oscilloscope screen could be used to gain more information than the echo trace on moving starch iodide paper.

He soon found that the naval scientists knew a lot about refraction, reflection and scattering of sound beams and the advantages that submarines might take of shadow zones. They had heard depth charges at ranges up to 100 miles in the Mediterranean and understood the deep sound channel. HMAS *Sydney*, using a more powerful sound source than had previously been available, had observed deep scattering layers on the way from Portsmouth to Gibraltar as early as 1932. Much of this knowledge might have been put to better use but the authorities, having had so much difficulty getting ASDICs (nowadays called sonar) into ships, seemed to think it better to keep quiet about natural limitations.

Deacon was married in May 1940 to Margaret Elsa Jeffries, elder daughter of Margaret and Charles Joseph Jeffries and sister of Sir Charles who was an Under Secretary in the Colonial Office. As the nation was at war seagoing became difficult: all the scientific equipment had to be taken out of HMS *Kingfisher* overnight so that she could go to Dunkirk to help evacuate the British army and after that there was more warlike activity than science. Even around Portland the war seemed close. On almost any day at sea there were fires along the coast resulting from enemy action and for fear of capture scientists had to carry a certificate saying that they were not taking part in the fighting. On one Sunday morning a bomb completely destroyed Deacon's office so that not a trace could be found of his work or his books including a number of the reports of the German Atlantic Expedition in the *Meteor*.

The anti-submarine research was moved in 1940 to an old-established boatyard at Fairlie on the coast of Ayrshire. The following year, the Deacons rented a house at Seamill on the estuary at West Kilbride and their daughter Margaret Brenda was born there in January 1942.

Some of the early NIO staff photographed at Teddington in Autumn 1952. (Left to right.) ***Back:*** *Norman Smith, Frank Pierce, Cyrl Williams, Rick Hubbard, D.W. 'Dick' Privett, Laurence Baxter, Leon Verra;* ***Front:*** *Jim Crease, M.J. 'Tom' Tucker, Henry Charnock, George Deacon, Ken Bowden, Jack Darbyshire. Note: Frank Pierce was absent that day so an assistant's body was borrowed to which Frank's (somewhat enlarged) head was attached.*

Fairlie was developing ahead-thrown weapons and compatible sonars. More attention was being given to the use of what had long been known of the refraction and reflection of sound beams, possibly helped by some feedback of interest from more academically promoted studies of anomalous propagation of radar. Deacon helped the staff of the Admiral commanding submarines with the first instruction book on the use of the bathythermograph in British submarines.

Deacon became a Fellow of the Royal Society in 1944. He believed, correctly, that his election to the Fellowship had been supported not only by marine physicists like Proudman, Doodson and Goldsbrough, but also by marine biologists in whose work he had always been interested.

At this time the Navy was becoming interested in oceanographic studies, particularly of sea waves. The superintendent, Dr A.B. Wood of the Admiralty Research Laboratory at Teddington, was concerned about the effect on pressure mines of waves and swell and that experimental evidence was lacking. Consequently, wave recording stations were set up on the south and north Cornish coasts to observe the period and height of waves coming up the channel from the Atlantic. A new unit, formally called the Oceanographic Research Unit but informally 'Group W', was therefore set up and Deacon was asked to lead it. He arrived there in early June "just before D-Day and Ocean Wave winning the Derby". He wrote that

The new oceanographic group was strong in individual talent; it is fairly certain that until then there had never been such a concentration of ability in physics, mathematics and engineering devoted to basic studies other than tides, and later on to those too. Among the group at the beginning or joining in the first year or two were N.F. Barber, K.F. Bowden, D.E. Cartwright, H. Charnock, J. Darbyshire, M.S. Longuet-Higgins, C.H. Mortimer, M.J. Tucker and F. Ursell, all to become well known for their contributions to marine science and to hold positions of responsibility; five of them were elected to the Fellowship of the Royal Society.

Just as important were F.E. Pierce, whose design and engineering ability brought a new dimension to oceanographic instruments, and N.D. Smith whose enthusiasm and ability enabled the development of many new methods and instruments. They were able to face the need to obtain detailed information on the rates of travel and decay of sea waves and to improve on existing attempts that used some kind of average of wave height and length, by treating the record as a mixture of sine waves and using spectrum analysis.

One of the outstanding advantages of the institute was its tradition for physicists, biologists, chemists, geologists and engineers to go to sea together. Many of the advances in equipment and methods could not have been made so promptly without such participation. This was true of cooperation with other laboratories, with defence laboratories, the National Physical Laboratory and the Post Office research laboratory, to name only a few. The scientists also gained inspiration and experience from direct contacts with their opposite numbers in industry and the professions. There seemed to be benefits all round; the scientist or engineer learnt what the industry was likely to need, and the industrialist what the scientist thought might become possible. When it came to getting something made or some special material a visit by the engineer or scientist speaking the same language and sharing experience with his opposite number in the factory was much more effective than an order office and tendering.

Deacon became Director of NIO in 1949. He recruited in a very informal manner, spotting talent where he could and appointing staff without the formalities of CVs, recruitment boards and interviews. In this way he built up a staff of enthusiastic and dedicated scientists and it was rare that his judgment was wrong. He directed the science in an equally informal manner allowing staff to pursue their own interests providing they fell within the general area of oceanography.

NIO continued to grow over the years at Witley and was given a new ship, RRS *Discovery*, to replace the aging RRS *Discovery II*. When the government decided in 1965 to put oceanography with other environmental subjects under the Natural Environment Research Council (NERC), Deacon was concerned that this would lead to over-centralised administration.

> *There was more money but less direct responsibility and involvement and this was a set back for a laboratory that had gained an international reputation for doing great things at relatively little cost, and later on as detailed accounting, administration and overheads added to the labour and cost of even minor transactions, exchanges with other laboratories and even between related projects in the institute itself were hindered by formalities, and it became much more difficult to share expensive facilities and hard won experience.*

After retirement in 1971, he was given a NERC Fellowship and an office where he continued to write papers on Antarctic oceanography, and its history. He also published measurements made in McMurdo Sound by Scott's expedition in 1911-12.

He returned to the Antarctic in 1975 in the US Coastguard icebreaker *Glacier*. They intended to study the expected flow of Antarctic bottom water down the continental slope east of the Antarctic Peninsula, but were unable to penetrate far enough into the ice. However, useful observations were made in the boundary region between the Drake Passage and Weddell Sea Currents. Early in 1976 on *Discovery's* 100th cruise, he returned to Antarctica to study the water movements and krill south of Africa.

Given a small retainer after his NERC Fellowship came to an end in 1979, he published more papers on the water movements and their apparent bearing on reproduction and distribution of krill. In 1981, he visited the Marine Studies Centre of the University of Sydney, Australia, on a Queen's Fellowship. He also wrote a book on the Antarctic circumpolar ocean dealing with the probable effect of the water movements on the life histories and distributions of the plankton.[6]

Deacon's many contributions to oceanography were recognised by the awards of a CBE in 1954 and a knighthood in 1971, when he retired as Director of NIO. But he also received many awards from academia in the course of his career (see Biographical Notes below).

He much enjoyed the presentation of the Albatross Award in 1982, a lighthearted but nevertheless serious award by international oceanographers for the "most obscure contribution to oceanography". The citation was "for fathering Margaret and the Institute of Oceanographic Sciences". The previous NIO recipient in 1960 was John Swallow,

John Swallow and George Deacon with the Albatross Award given to Deacon in 1982. It had been awarded to John Swallow in 1960.

"For innovative measurements of ocean currents both AC and DC. The deep circulation has never been the same since Swallow figured out a way to measure it directly." The stuffed albatross in its cage had to be presented during a conference held abroad and brought back to the recipient's country with its attendant literature to persuade customs officials to allow it in. Today this migration is made more difficult by the CITES (Convention on International Trade in Endangered Species)!

Deacon died in November 1984. In the Biographical Memoirs of Fellows of the Royal Society,[7] Henry Charnock, Deacon's successor as director of NIO, wrote that

> *... he was much respected by his younger colleagues, who always addressed him as Dr Deacon: only his old Discovery Investigations co-workers would presume to call him by his first name. There was an occasion when one of his Group W, thinking his group leader miles away, referred to him as 'old George' only to find Deacon by his side. Very embarrassed, he muttered an apology. 'I hope you don't mind my calling you George'. 'Not at all' replied Deacon, adding, after what seemed a long few seconds, 'My wife calls me George'. It was probably not intended as a rebuke–his parents and sister still called him Ted–but Dr Deacon he remained thereafter. It is not given to many to found and to form an institute that so rapidly acquires an international reputation: Deacon had a young but enthusiastic staff and he encouraged them to seize their opportunities.*

Deacon's staff admired and respected him, both for the freedom that he gave them in their research and for the wisdom he showed in encouraging new adventures in science and in maintaining a wide spectrum of disciplines in oceanography. This proved to be hugely beneficial when NIO addressed some important practical problems such as the disposal of radioactive waste, or the interaction of the oceans and atmosphere.

He was always very approachable and was modest about his own achievements. It was not always obvious to his staff how much he fought behind the scenes for the resources, both manpower and financial, to achieve his vision of the science of oceanography. His lasting achievement is the expansion of oceanography not only at NIO but in the UK and internationally, and also in the laboratories that followed NIO: the Institute of Oceanographic Sciences (IOS), the Witley laboratory which was later named the IOS Deacon Laboratory in his honour, the Southampton Oceanography Centre and the National Oceanography Centre, Southampton. His scientific legacy is the basis of modern oceanography in the UK today.

Achievements and awards

Deacon was elected FRS in 1944, FRSE in 1957, awarded a CBE in 1954 and knighted in 1971. He was awarded Honorary Degrees at the Universities of Liverpool and Leicester, was made a Foreign Member of the Swedish Royal Academy and an Honorary Member of the Royal Society of New Zealand. He received the Polar Medal (1942), the Alexander Agassiz Medal of the US Academy of Sciences (1962), the Royal Medal of the Royal Society (1969), the Albert Memorial Medal of L'Institut Océanographique, Monaco (1970), Founder Medal of the Royal Geographical Society (1971), Medal of the Scottish Geographical Society (1972) and the Albatross Award of the American Miscellaneous Society (1982). He served on numerous committees of the Royal Society and of several marine, meteorological and polar organisations, both national and international. He was the author of many scientific papers and author or editor of several books.

4

Group W at the Admiralty Research Laboratory

Personal reminiscences by some of its members

Compiled by Michael Longuet-Higgins

Fritz Ursell[1]

The Admiralty Wave Group (Group W) was set up on 5 June 1944 at the Admiralty Research Laboratory (ARL) in Teddington, southwest of London. The invasion of Normandy took place one day later, on D-Day, 6 June. For this invasion a joint Anglo-American team had made accurate wave forecasts. They had found that the waves on the Normandy beaches would be very high, and for this reason General Eisenhower had postponed the landing by one day. The waves continued to be high, but a further postponement was not feasible. For the purposes of wave prediction the Normandy beaches formed one side of an inland sea. The surf incident on these beaches had, in effect, been generated by local winds; it was not due to waves propagating from the North Atlantic. The forecasts made use of correlations which had been established between local wind and local waves.

It was realised by the military planners that at a later stage of the war (say in 1946) there would probably be landings from the Pacific Ocean, perhaps in the Philippines or in Japan. It was understood that in the Pacific the relevant waves on beaches would probably not be local but would be swell generated by storms that had occurred many days before the landing and at a distance of many thousands of kilometers from the beach. Thus the rules for wave forecasting in the Pacific would need to be different and would also need to be established in good time. Group W was set up to establish them. No one had any idea what these rules would be. Since the results were not wanted before 1946 the project had zero priority; we were not permitted to ask for a ship to put down equipment in the sea, but we were permitted to use equipment that was already in the sea and was no longer needed for its original purpose. (It turned out that this was sufficient). The security grading was also very low.

In June 1944 there was no unused reservoir of scientists. The number of scientists needed for the new Group W was small (about six) and was found largely from ARL Group H (Electromagnetics) which had come to the end of its highly successful degaussing programme. As may be imagined, the people who were transferred were the ones who could most easily be spared. To put it more bluntly, all of us were rejects of one kind or another. I came to join the group in the following way. I had joined the Admiralty Department of Scientific Research and Experiment in December 1943 after graduating in mathematics at Cambridge University and had become a member of Group H. The head of that group was Dr S. Butterworth, a very distinguished physicist and a kind chief. He did not really need the type of assistance that I could give him, and I worked on odd problems as they arose. I had somehow acquired a large desk which was to be moved for me to another room to make way for the new Group W. The desk was too big for the other room. Dr Butterworth suggested that I stay put with my desk. I did so, and thus the whole of my subsequent career was determined.

The head of the new Group W was Dr G.E.R. Deacon, DSc, FRS, who came to Teddington from anti-submarine work in Scotland. As a marine hydrologist of great distinction he should, I suppose, have been in charge of a larger and more important group. Perhaps in his previous work he had pointed out some unwelcome truth. We came to know him as a person of great integrity. Of great importance to me was Norman Barber, MSc (Leeds), who had joined the Scientific Civil Service before the war. Then there was Jack Darbyshire from North Wales, with a first-class degree in physics from the University of Wales, Bangor. I must also mention Dr C.H. Mortimer, a freshwater biologist with theoretical and experimental skills, also M.J. Tucker, another physicist of ability. These and other colleagues formed a distinguished group of rejects.

We soon moved to another room, a large glass conservatory over the ARL workshop. We also had a laboratory downstairs. The conservatory had great advantages, but the glass had potential disadvantages when the V-1 flying bombs started to rain down, a few days after D-Day. We could not know that the glass was to survive these attacks, unlike much of the glass in Teddington. It was Dr Deacon's idea that there should be no partitions in the conservatory; his own desk was at one end, then there were our desks, and the drawing-boards of the draughtsmen-designers were at the other end. We were all of us visible and audible to each other. There was thus no obstacle to discussion. This was good, as we did not know how to approach our problem. To begin with, the waves we needed to measure, in any experiment that we might undertake, had to come from a great distance, in other words from the North Atlantic. This ruled out most of the British coastline and left only the Outer Hebrides in Scotland and North Cornwall in England. We chose North Cornwall. What measurements would be useful for our purpose? We spent some weeks in Cornwall, seeking inspiration, but none came, and we returned to Teddington.

I was the only member of Group W with sufficient mathematics to read Lamb's 1932[2] book on Hydrodynamics. This great work contains two chapters on water waves, one dealing with tidal waves and the other with the kind of waves generated by storms. Here I found a full account of the problem of Cauchy and Poisson who quite independently in 1815 found that if one observes waves at a great distance from their source the wave period (the time between successive waves) is constant at a point which moves away from source with a velocity called the group-velocity. In deep water this velocity is one-half of the velocity with which the wave crests are seen to travel (known as the phase velocity). This means that if one observes the wave period at any given time, then one can trace this period back to its source with the appropriate group velocity. For a small localised distant storm the observer would see a single wave period decreasing with time and for a more complicated wave source (a storm spread out over an area and over a period of time) there would be a spectrum of periods, again with the longest (lowest frequency waves) arriving first.

Based on this theory I suggested that we should measure the variation of frequency with time in Cornwall and attempt to trace the waves back in space and time and see whether they did indeed originate in an area of high wind. (There was plenty of weather information.) I wrote a short note about this in 1945.[3] This was a rather theoretical suggestion. Observation showed that there were local waves as well as swell, and that the local waves were visually more prominent than the swell. In fact we were able to measure bottom pressure, which emphasised the longer waves, but no obvious periods could be seen. What was needed was a frequency analysis, but this would not be useful for our purpose unless it could be done in real time. It was decided to make 20-minute records of wave pressure at intervals of two hours.

The wave analyser.[4] *The black and white wave record can be seen on the drum.*

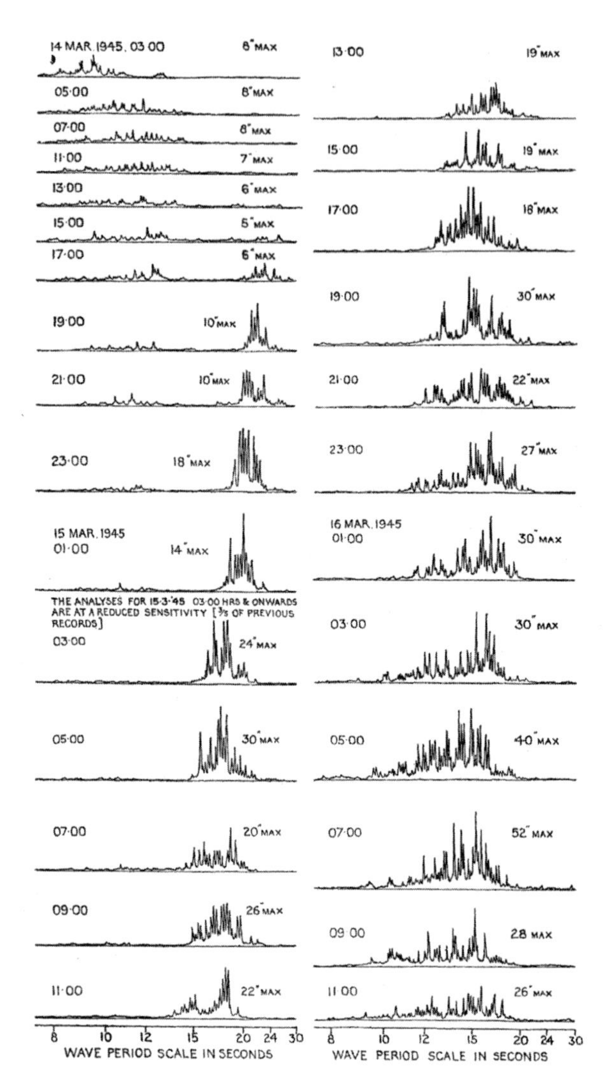

Wave spectra at 20-minute intervals (from Barber, N.F., and Ursell, F. 1948).[5]

Each record would also record a time signal at 20-second intervals; the simultaneous (Fourier) analysis of this would give a calibration of frequencies in the pressure record.

The wave record was a black-and-white profile which could be read optically by reflection. The record (of length approximately 20 minutes) could be made to fit exactly on to the circumference of a large wheel. There would of course be a jump in the recorded pressure amplitude between the end and the beginning of the record. The record would be repeated over and over again, by turning the wheel at say 4 revolutions a second. The periods in the record were therefore known, they were the harmonics of the length of the record, i.e. 8 times, 12 times, 16 times... a second.

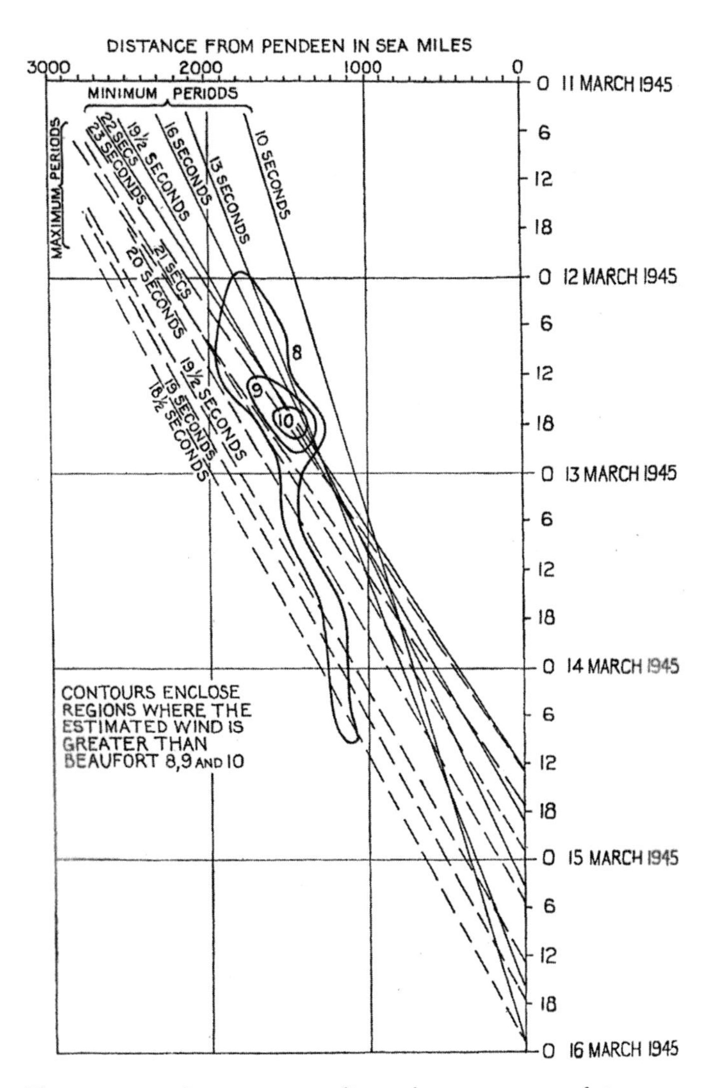

The propagation diagram corresponding to the wave spectra on facing page.

For each period there would be a tuned circuit, i.e. 200 circuits for 200 harmonics. The amplitude of each harmonic could therefore in principle be measured by recording the current in the appropriate circuit. (The jump between the end and the beginning of the record would contribute a wave of period 20 minutes, outside the range of ocean waves.) This method would have been very laborious and rather sensitive. If the speed of rotation of the wheel was not kept absolutely constant during the analysis, then all the amplitudes would be knocked back to zero every time the join came round, four times a second. If the speed of rotation was not the same from one analysis to the next, then all the circuits would need to be retuned.

These difficulties were overcome by Norman Barber in masterly fashion, and the success of the experiment is largely due to him. The wheel, a heavy wheel with nearly frictionless bearings, was speeded up to four revolutions a second and the speed was then left to slow down under friction.

There was a single tuned circuit that read the record optically. As the speed decayed, the frequencies successively came into resonance with the tuned circuit.[4] No effort was required, and the analysis took about 20 minutes. Records were taken at intervals of two hours. As we had expected, the spectra move to the left. They are of finite width: long waves from the later part of the storm arrive at the same time as short waves from the earlier part of the storm. Wind strengths are entered on the propagation diagram.

It will be seen that propagation lines from the high and low end of the spectrum pass through areas of strong wind in just the way that would be expected. Many more examples are given in the original paper by Barber and Ursell.[5] The conclusion is that outside the storm area the waves propagate in accordance with the linear theory. This is perhaps a surprising result that was obtained as early as 1945. The next step was to correlate the waves propagating from the storm with the strength and direction of the wind inside the storm area. This task was immediately undertaken by Jack Darbyshire[6] with considerable success, and his pioneering work was imitated and improved by others.

The war in Europe ended in 1945 and since our work was no longer needed for military purposes we were permitted to publish it. A draft was prepared, mainly by Norman Barber. This was completely rewritten by George Deacon and Michael Longuet-Higgins.[6] According to Admiralty rules, Deacon should have been the sole author. Actually his name appears only in the phrase "Communicated by G.E.R. Deacon, FRS". The work attracted a great deal of attention. I remember visits to Group W from many distinguished geophysicists, and in particular a visit by H.U. Sverdrup and Walter Munk. Visitors were always received by George Deacon in our glass conservatory. He would describe our work very clearly, and his lecture would always end in the same way. He would assert that he had of course been in administrative charge, but actually the work had been done by … (and then he would give all our names.) He did the same in print, see for example his lecture to the Challenger Society in 1946, reprinted in Margaret Deacon's book.[7] I now realise that he had contributed much to our success. I am sure that he must have been under pressure to produce results, publish interim notes etc., and that he used his prestige to give us freedom from the administrators. Perhaps he also contributed in other ways which we were not experienced enough to appreciate. (In 1944 I was 21 years old.) He was determined not to claim any credit to which he was not justly entitled, and he therefore renounced all credit. When the National Institute of Oceanography was founded soon afterwards he was appointed its first Director. He and I were never close to each other but he was to become an inspiration to me. As for Norman Barber, he was a true physicist, with physical insights which (as I then discovered) differ greatly from mathematical insights. When we were using Fourier analysis he wished to find a physical interpretation of Fourier's Theorem and ultimately wrote a little book about it. At first I was astonished, since to a mathematician like myself the mathematical proof was enough, but I began to see that there could be other valid approaches. Of all my colleagues he had the most profound influence on me. For me he was the closest approximation to a thesis supervisor, for I never was a graduate student. We remained in touch until his death in 1992.

(Left to right) Fritz Ursell, Michael Longuet-Higgins and Clifford Mortimer aboard a launch on Lake Windermere, on the occasion of an international limnological conference in 1949.

Jack Darbyshire[8]

In January 1943, three and a half years after the outbreak of the Second World War, I joined the Admiralty Research Laboratory, Teddington, where I was assigned to Group H under Stephen Butterworth. The 'H' stood for hydrodynamics, but in my time hardly any work was done in that subject, and most effort was directed at the magnetic detection of submarines and similar objects. However, by 1944, it had become vital to study the ocean for its own sake, particularly the action and formation of sea waves.

While in Group H, my co-worker, Norman Barber and I were concerned with trying to detect rather small 'iron objects' which could be sent into harbours. To do this we had big wire loops on the sea bottom and the magnetic effect of the 'objects' would send a current through each loop which we could detect. Somebody had the brilliant idea that one could measure waves by this method, i.e. if you had an iron buoy moving up and down, then you could measure the height of the waves by its up and down motion. This was a relevant problem at that time, as the invasion of France was imminent. It was therefore important to be able to measure the waves in the English Channel and correlate their properties with the wind. So we got involved with the problem, and it was decided to split us from Group H to form a separate group, which was eventually called Group W–'W' for waves.

It turned out that there were several wave recorders placed along the south coast from Dover to Penzance, and a few on the north coast of Devon and Cornwall, measuring variations in pressure at the seabed, and it had been decided to attempt to correlate these wave records with the wind, which could be easily measured. This work was already in hand when we became involved. Group W consisted of Barber (who was then its head) myself and two assistants, Hubbard and Grey. There were also two Scientific Officers, Cotsworth and Alexander, from the Mine Design Department in Havant, who were concerned with servicing and looking after the wave pressure recorders along the coast. There was also Clifford Mortimer, a freshwater biologist who before the war had been working at Ambleside, near Lake Windermere.

Steven Butterworth, whom I mentioned earlier, was a great mathematician and famous for his theory of the transformer. He had a mathematical assistant who was called Wigglesworth (a good name for a mathematician) but Wigglesworth was transferred to Scotland and Butterworth was extremely annoyed, so they gave him a chap called Fritz Ursell who had just graduated in mathematics with high honours at Trinity College, Cambridge. He was a German refugee who had come to Britain before the war and had been educated at Marlborough before going to Trinity. It so happened that his office and Butterworth's were next to ours–and so a certain amount of 'leakage' took place. Naturally, when Barber and I started on the theoretical work, we went to ask Ursell for his advice, and so he gradually became more interested in our work than the work he was supposed to be doing. After a while he got himself transferred to our group and he was, of course, a great asset.

Barber was a newly promoted Senior Scientific Officer and was not really senior enough to head a group so they found an eminent oceanographer–George Deacon, who had just been awarded an FRS, having written an important paper on the circulation of the Southern Ocean. At that time, Deacon was working at the anti-submarine establishment at Fairlie in Ayrshire. It so happened that the work he was doing was of far more use to submariners than anti-submariners. As a result, although the submariners were duly grateful, he was not very popular with the others, and so he was not averse to transferring to Teddington. He did not know much about wave theory (he had been trained as a chemist) but as he had circumnavigated the Southern Ocean several times, he knew all about real waves.

Deacon packed us off to the Cornish coast where there were some wave recorders in action. Barber, myself, Ursell and Grey travelled to Padstow along the A30, in a V8 Landrover [*Probably some other 4-wheel drive vehicle since the Landrover did not exist then. Eds.*] driven by Alexander. At Bodmin we had to interrupt our journey because Ursell was technically still an enemy alien. He had to register with the police, so we stopped outside the police station, which was the County Headquarters for Cornwall, and we went in. Some time later a policeman came out and asked us if we knew a Sergeant Clerk in Teddington, which was obviously a ploy, for if we had said 'yes', they would have thought that we were enemy aliens or something similar. In fact the Cornish Police were the nearest I have ever seen to the Keystone Kops.

It soon became clear that there was no regularity about the waves, and that the only way to deal with them was to make a kind of histogram. The lengths of wave traces on the record could be split into different groups and the number of waves in each group counted. So you could plot a frequency distribution and for each particular group you could work out a kind of mean amplitude. This was a laborious process but it seemed the only way forward.

There were some amusing instances in Padstow. Constantine Bay was three miles away and we did not have use of the car (and we could not drive anyway) but we had one bicycle. So George Grey and I travelled there, two on a bicycle. But in Padstow we were stopped by a war reserve policeman, who also happened to be the local grocer. He asked our names and addresses. Because my address was exactly the same as the one that Fritz Ursell had already given the police, and because of my accent [*Darbyshire had a very pronounced North Wales accent. Eds.*], the 'grocer policeman' thought I was the German refugee and that I had given a false name. When Ursell saw the real policeman he sorted it out. The 'grocer policeman' must have been very slow, because I found later on that there was a gang of quarry workers and their wives in Padstow and, as they came from Trefor in North Wales, their accents were exactly like mine!

After about two months we felt we had done all we could in Padstow, and we went back to Teddington to face the flying bombs and the V2 rockets once more. By this time we had been given new quarters in Teddington. These had previously been the shipyard and the shipyard drawing office. In the shipyard they made models of warships with steel or iron so that their magnetism could be studied. The drawing office, where they had designed the models, became an office, and the shipyard our lab. The workshop itself was very good, with plenty of space. There was a gallery on one side where the remnants of the shipyard workers were still working. The offices were all on the roof, in a prefabricated building.

There was plenty of space and, on the face of it it was ideal. However, the office was very cold in winter, and not long after we had to face probably the worst winter of the 20th century, the winter of 1946-47.

We then decided to do something better than the histogram, although along the same lines. The original idea was to send the wave record through a lot of different filters, one filter for each period, but this was not really practicable. George Deacon had a friend in the film industry and we learned that the Walt Disney film 'Fantasia', which was then a recent success, had the sound part of the film as black and white wavy silhouettes along the side of the picture frames. It therefore occurred to us that we could do the same thing, only using not transmission but reflection. We could use the type of paper we used in Padstow, and instead of printing lines we would have a big block of light which would move about, generating a silhouette of the waves. We could then put this on a wheel and, as the wheel spun round, the variations in black and white could be detected by a photocell.[9]

By this time, most of the recorders had been discontinued but there was still one at Pendeen. We had a man looking after it who worked for Western Union and he used to send us terse messages with the records. He would send them undeveloped and we would develop them in Teddington. Then, having dried them, we put them on the wheel. We started analysing on a regular basis. The records were taken every two hours, so some days we had as many as twelve to process.

We pressed on with the analyses. At first they all looked very complicated, a mass of lines, but one day the analyser record showed a narrow band at about 20 seconds period. As we analysed more of the records, this narrow band gradually decreased in period, getting shorter and shorter. At long last we realized we were on to something significant. In the storm areas there would be a mixture of waves with different periods. Longer period waves travel faster than those of shorter period, so they would arrive first and would be followed by the shorter period ones. Hence the waves were dispersive, and the period of the peak in the spectrum would get shorter over time.

This was quite an achievement: we had never seen anything like it before, and we called it 'Exhibit A'. A week or two later there came an even better example. This one lasted for over a day, so long that we were able to identify, by working back, where the waves came from. We were using six-hourly weather charts of the Atlantic from the Naval Weather Service, and when we plotted on a time/distance graph the wind speeds pointing towards Cornwall, then worked back, we could identify the sea areas where the waves had come from. Barber found that numerically the wind speed in the storm in knots was twice the period of the wave in seconds; in other words, a 20-knot wind would cause 10-second waves.

This looked rather suggestive because the speed of the surface wind, which was estimated at two-thirds of the speed of the geostrophic wind, would in this case be the same as the wave speed.

At about the time we finished the analyser, M.J. Tucker (we called him 'Tom' Tucker) joined us. He was a proper electronics expert. He started building a new analyser based on a tuning fork filter rather than on a vibration galvanometer. His analyser was transferred to our station in Perranporth, which I will describe later. Our wave analyser was more or less 'a nine-day wonder', as in those times it would take a human calculator a week to analyse a five-minute record, whereas our analyser could do a 20-minute record in 15 minutes. A lot of important people came to see it, including Sir Henry Tizard, the Chief Government Scientist, Professor Bernal, and Sir Frederick Brundrett, the Admiralty Director of Scientific Research; also Professor Sverdrup and many others. In fact, at that time, we were the only people in the world as far as I am aware, who could make Fourier analyses of long records.

The Americans had been successful in predicting the swell for the North African landings in 1942, using methods devised by Sverdrup and Munk. I am inclined to think that they were rather lucky because their method was based on very old visual observations. They had nothing to compare with the results that had been obtained by our group.

From our graphs we could now work out the distance over which a wind of given speed had acted on a wave of chosen period, and for how long. It was assumed throughout, however, that the waves did not interact with each other.

The war ended rather sooner than expected, in August 1945. I was sent to measure the waves at Perranporth (a new site, not one of those used for studying waves at the time of the invasion). We installed a wave recorder there and we built a substantial hut, in which the analyser designed by Tucker was set up. We therefore had dark room facilities for developing the records, and of course the means to analyse them.

During the experiment I went down to the beach at Perranporth to see what was happening. It was very strange – the sea seemed remarkably calm but not quite flat, and there was a peculiar swell. I started analysing the records. It was very convenient having everything at hand – wave recorders, the analysers and development facilities, etc. – so I started processing the records every two hours, not every four, which was the norm. Soon it became clear that something peculiar was happening. There was a very narrow band of frequencies corresponding to a very long period of 24 s and there was nothing else on the records, and the period gradually shortened as time went on – this lasted for several days, with the period decreasing only very slowly. I made a rough estimate of the distance from the storm and it came to about 7,000 miles. However, it turned out that there was a strong tidal influence, a kind of Doppler effect, in which the ebb and flood of the tidal currents caused the wave period to increase or decrease – the same kind of effect as when a police siren passes the listener. I analysed all of these records and took them back to Teddington.

When the whole dataset were plotted out and the Doppler effect was allowed for, the storm generation region seemed to be about 10,000 miles away near Cape Horn. Nobody had ever tracked waves travelling over such a long distance before.

Deacon had wanted Barber and Ursell to write and present a paper about the work, possibly at the Royal Society. But the only discovery was that under certain conditions waves travelled with the theoretical group velocity and this was hardly enough to make a paper. So the result from Perranporth proved to be a godsend to them because, not only was it a much better example of waves travelling with the theoretical group velocity, but also we had observations of the Doppler oscillations due to the tide. In fact, this effect is not quite so simple as the police siren case. They made rather a meal of it, and Barber was to write a separate paper on that alone.[10] Anyway, the paper went ahead and eventually was printed in the *Philosophical Transactions*.

Now I must mention the subject of microseisms. These are very tiny sea-bed oscillations which are supposed to be excited in stormy areas and then propagate outwards. There were seismographs at Kew Observatory – only a fourpenny bus ride from Teddington – which provided us with access to microseism data. The data were usually recorded graphically on a large sheet of paper, about half a metre square. The ground movement was denoted by wavy lines, one for each hour.

It was fairly easy to work with amplitudes and Deacon wrote a paper in 1949[11] which showed a correspondence between the amplitude of the microseisms at Kew and the wave heights at Perranporth. There really was some sort of relationship, but then the problem was to derive the spectra of the signals. This was not very easy because from the graphical lines we had to somehow obtain the black and white silhouette form we were used to. There were basically two methods. The hard way was to photostat the original record and put the resulting strip in a photographic enlarger and magnify it to a respectable size. The profile was then traced from the enlargement onto paper the same size as the photographic paper we used with waves. The easier way was to use a photo-electric curve-follower which had been designed very cleverly by Tucker and Collins – basically, it was a narrow beam of light that automatically followed the record.

One way or another, we did obtain many examples of interesting wave spectra from Perranporth and checked them against the microseisms recorded at Kew.[12] Not only was there a very strong correspondence between them, but the microseism spectra preceded the waves, although not by very much – only a matter of about six hours. So they could not be actually related to the centre of the storm, but were instead probably due to reflections off coasts nearer to the storm area. But there was one example, in November 1945, when there was a severe gale and the microseisms provided a 24-hour warning. In that case, when we checked the meteorological data, we found that the winds had in fact veered around very quickly. This event thus provided clear evidence that an increase in microseism energy at the coast could be used to monitor wave activity at the centre of a storm.

Jack Darbyshire after his retirement as Professor of Oceanography at UCNW Bangor in 1986. He was born in 1919 in Blaenau Ffestiniog in North Wales and died in 2005.

Clifford Mortimer[13]

Trained as a biologist (University of Manchester 1924-32) and with ambitions for a career in genetics, I chose a leading centre in that subject for postgraduate training, the (now) Max Planck Institute for Biology in Berlin. Returning to Britain in 1935, no jobs in genetics could be found, and so I joined the recently formed Freshwater Biological Association (FBA) as a research student at £150 p.a. at Wray Castle, English Lake District. That small salary was accompanied by a gift of much greater value, the freedom to follow where my curiosity might lead. The only stipulation required me to continue W.H. Pearsall's water chemistry analyses of the lakes.

As a Quaker, when war was declared in 1939, I was faced with a difficult choice: (i) conscientious objection; (ii) the Quaker Ambulance Service; or (iii) registration in C.P. Snow's list of scientists available for the war effort. Because of my close observation (in Berlin) of the evil Hitler gang, I felt I must do all I could be prevent their takeover of Britain and Europe. So I registered; wrote up my findings in lakes; and waited. After one rejection, probably because my wife Ingeborg, a British citizen by marriage, came from Germany, a call came to join (as a civilian scientist) the Admiralty's Mine Design Department (MDD), based in Portsmouth Dockyard on the south coast. Because of frequent bombing, MDD had moved to a nearby country house, West Leigh in Havant.

In June 1944, a selected group of scientists and engineers were transferred from various Admiralty groups to a new Group W to be housed in the Admiralty Research Laboratory (ARL), at Teddington in West London. Norman Barber and others named later came from ARL. Cotsworth and I came from MDD; and George Deacon, an experienced oceanographer and our director, came from an antisubmarine unit at Fairlie in Scotland. We all occupied one large room at ARL; and Inge and I moved into another wartime-evacuated house close to the river with a bomb shelter in the garden. Our small group of physicists, mathematicians and engineers worked remarkably well together. Cotsworth and I were mainly employed in setting up and operating wave recording stations.

My clearest memory of the stimulating time in Group W was of Norman Barber's intuitive grasp of the physics and theory governing the phenomena we were investigating. He also had the gift of communicating the physical picture to those who, like myself, were not fluent in the language of mathematics.[14] I recall watching, with him, waves arriving on a Cornish beach and hearing him ask: "How can we disentangle the wealth of information carried by these waves coming in from local wind action and from distant storms?" The obvious answer was spectral analysis, but "how could this be performed quickly enough to provide forecasts of the timing and force of storm arrivals at the measuring site?" Today this can be done with computers and Fast Fourier Transforms, not available then.

A description of the wave analyser is given elsewhere in this chapter.[4] Cotsworth and I routinely recorded the waves, for 20 minutes every two hours, as black and white silhouettes on photographic paper strips, the length of which equalled the circumference of a large wheel of the kind used with belt-driven machinery in factories.

First, we located a long straight beach with good ocean exposure at Perranporth on the North coast of Cornwall. There, we deployed, on the seabed, robust pressure sensors, connected by armoured cable, to a recording hut on a cliff overlooking the beach.

A series of spectra, obtained at Perranporth by the method just described at two-hourly intervals, 14-16 March 1945 enabled the waves to be traced back to a storm off Newfoundland.

Measurement of wave heights and periods on repeated spectra, coupled with theory, permitted forecasts to be made of time of arrival and wave height of storms. That combination of theory, instrumentation and monitoring was characteristic of Group W's collaborative achievements.

Cotsworth and I were the 'meteorologists' mentioned in Rachel Carson's account,[15] in colourful prose, of the Perranporth endeavour. American oceanographers and naval personnel were also interested in our work. I recall a visit to our recording hut by H.U. Sverdrup, director of Scripps Institution of Oceanography in La Jolla, California. Strolling along the cliff–top path one wind-free evening, he was smoking and blew a perfect smoke ring from his mouth. Leaning forward, he inhaled; and

The observing site at Perranporth, Cornwall. The wave recording station (upper building) had a large plate-glass window looking north along the 3km-long beach to allow observation and photography of the waves. The small lower hut belonged to the Coast-Guard.

it disappeared into his nose. A few seconds later, a perfect ring emerged from his mouth, demonstrating (he said) the principle of conservation of vorticity.

'Tom' Tucker

During the war you had to get your degree before your 20th birthday or you were called up into the armed services, so I qualified in 1944 with a degree in physics with subsidiary subjects maths and chemistry. When you had received your exam results, you were interviewed and told what job to go to. I was sent to the Admiralty Fuel Experimental Station, which consisted of a brick-built laboratory and a Nissen hut as an admin office on the banks of a dismal muddy creek in Haslar, near Gosport. A captain was nominally in charge, but we practically never saw him. Otherwise there were two chemists, a sailor as a lab assistant and a WREN as the captain's secretary. They had asked for a physicist, but had no idea what they wanted him to do, and as a newly qualified physicist aged 19 years, I was not able to find myself a useful physics job. So I used my chemical knowledge and spent my time there analysing fuel oil samples. As you might imagine, I was not happy and soon asked for a transfer. I then fell on my feet and in December 1944 was transferred to the newly formed Group W at the Admiralty Research Laboratory at Teddington. I was happy there and stayed with it and its successors throughout the rest of my career.

I turned to electronic engineering as my speciality. There was plenty of scope for this in Group W. However, my degree course had not included this subject, so in the early days I was learning my craft. Fortunately, there were two experienced practical electronic engineers in the group and one of these, Geoff Collins, who stayed with us for many years, was particularly helpful. I also made a close friend of one of the electronic engineers in the Admiralty Gunnery Establishment, who shared the site with us. I cut my teeth designing the electronics for the wheel analyser, and later moved on to designing new types of wave recorder and gadgets for analysing wave records. Much of this work was done in co-operation with Frank Pierce, our very able mechanical engineer. In fact, this co-operation, which developed into wider co-operation between the Applied Physics and the Mechanical Engineering Groups of NIO, lasted throughout the life of NIO, and was happy and fruitful. Dr Deacon was a marvellous boss and gave us consistent support in this work.

At ARL we had the support of Dr A.B. Wood, its Superintendent, who was well known as an expert on underwater sound and as the author of a classic book on the subject. In 1946 he left ARL to become Deputy Director of Physical Research for the Admiralty, where he was able to continue his support at a higher level. On his retirement in 1950 he came back to ARL to continue his research in underwater acoustics. He often visited Group W and gave us helpful advice.

The first wave recorder available to Group W had recently been installed at Pendeen in Cornwall off a westward-facing beach. It was an American Powerphone which measured the pressure on the seabed in a depth of 110 ft (approximately 33 m). The output was recorded by a mirror galvanometer connected to it by an underwater cable. The pressure on the seabed due to waves drops as the water gets deeper, and is down to half that corresponding to the surface amplitude at a depth of approximately 1/5 of a wavelength. Thus, at a depth of 33 m one is only effectively measuring waves with a length greater than 170 m, corresponding to a wave period of slightly over 10 seconds. This is fine for long swell, but is not adequate for measuring most storm waves, so in September 1945 another installation was made at Perranporth, again in Cornwall, where there is a long beach facing west. This used two pressure recorders made by the Cambridge Instrument Company mounted on tripods, one in approximately 13 m of water and the other in 22 m, connected to shore by an armoured cable. On each tripod there was also an upwards-looking echo sounder, intended to measure the surface elevation. The outputs from these instruments were recorded in a hut built high on the cliff overlooking the beach.

I was responsible for maintaining this installation for most of its life, which was a job that I really enjoyed. To get to it by train one passed through stations with names like Goonhavern and Goonbell Halt. The day-to-day operation was in the care of a scientific assistant called Edgar Gubbins. He was a charming man who had been a wool merchant. He had made enough money to retire early before the war and had built three (if I remember correctly) houses in Perranporth, and was still living with his wife in the last of these. He was big, taking size 14 shoes and with hands as large as spades, but he handled the rather fiddly pen recorders with great delicacy and took an interest in what we were doing. He told me that when he enlisted in the army during World War I, they had no boots big enough for him, so he had to spend the first 17 days in bed until they got a pair made!

In practice the echo sounders proved to be useless because in a storm the breaking waves filled the top of the water column with air bubbles, blanking out the echoes from the sea surface. It is interesting that at regular intervals since, people have again tried upwards-looking echo sounders as wave recorders with the same result. The experience does not seem to have entered into folk-lore!

The Cambridge recorders were successful, but after about 18 months they failed due to the air in the rubber bellows, that transferred the water pressure to air pressure, leaking out. We had to develop a more reliable instrument. It so happened that another department in the Admiralty had a large number of 2.5 inch (6.25 cm) diameter quartz discs 1/8 inch (3 mm) thick going spare. Each was cut from a single natural crystal and would have cost a fortune to manufacture specially. Quartz is, of course, a piezoelectric material and when squeezed these crystals produce an electric charge. We designed what was in effect a cylinder of 20 cm internal diameter (if I remember correctly) with a piston at one end pressing on a pile of six of these crystals arranged with alternate polarities and separated by steel plates. Quite large charges were produced by the fluctuating wave pressures. We had to hold these without appreciable leakage for a wave period and measure them. To do this we collected them on a capacitor and measured the voltage with an electrometer valve. We arranged it so that we could feed a step voltage down a separate wire in the cable to the 'earth' end of the capacitor, when required, to check the calibration and the leakage rate of the electric charge. These 'Piezoelectric wave recorders' proved to be successful and reliable.[16]

Soon after the end of the war we began to be consulted by coastal engineers wanting information about waves in connection with coastal engineering works, such as harbour construction. In fact, it slowly became obvious that the main practical application of our wave forecasting skills and of our general understanding of waves was not in real-time wave prediction, but in determining wave climate. The most important aspect of this was usually predicting the extreme wave conditions likely to be experienced at a site. Subsequently the research started in Group W has found applications in connection with the design of offshore structures such as gas and oil platforms, and in the generation of power from waves. Indeed in 1953, we were visited by a young engineer studying the feasibility of this on the west coast of the UK. I said to him that the sea is a hostile place and that any device would have to be strong and heavy. He replied "steel is cheap"! More than 50 years later we in the UK at last seem to be on the verge of producing commercially viable installations.

Michael Longuet-Higgins

September 1945. My first job. World War II was thankfully over, but for many scientists and mathematicians like myself there lay ahead three years' 'work of national importance'. After a two-month wait I was directed, as one possible choice, to the Admiralty Research Laboratory (ARL) in Teddington, Middlesex, next to Bushy Park. Entering through a gateway topped with broken glass and barbed wire, I was escorted up an iron staircase to a greenhouse structure perched on the roof of an engineering workshop. There to greet me was a cheerful, burly figure in shirt sleeves, George Deacon, Head of Group W, which was formed on 5 June 1944, the day before the landings in Normandy. The purpose of Group W, as Deacon explained, had been to study and forecast ocean swell for future military operations in the Pacific, but it had a wider remit: to improve our knowledge of the environment in which the navy operated.

I was introduced to members of the Group: N.F. (Norman) Barber, a tall, modest, ex-teacher from Yorkshire, who during the war had done vital work on the degaussing of ships (to prevent their setting off magnetic mines); Clifford Mortimer, a cheerful, energetic, business-like Quaker from Manchester, who before the war had studied biology in Germany (and had a German wife); F.J. (Fritz) Ursell, a Cambridge mathematician like myself, but two years ahead, he was a pre-war refugee from Nazi Germany, and we were to become close friends; M.J. 'Tom' Tucker, an able young physicist from London University; J. (Jack) Darbyshire a physics graduate from the University College of North Wales at Bangor, somewhat wild and with a strong Welsh accent; F.E. (Frank) Pierce, a dark-haired, bushy-eyebrowed marine engineer, who had also worked for the GEC; G. Collins, a small genius of an instrument-maker, who had previously designed optical instruments; and Norman Smith, a quiet man, who had flown over Germany many times during the war and since then had gained expertise as an instrument maker for Dr Cooke, a former Chief Scientist at ARL. In addition to this core group, there were several other supporting staff.

Dr Deacon described to me some of Group W's achievements. What especially impressed and attracted me was the wheel harmonic analyser invented by Norman Barber. With this device they had been able to distinguish ocean swell in Cornwall as having originated from high winds several thousand miles away, even as far as Cape Horn. They could also observe fluctuations in the wave period caused by tidal streams over the continental shelf west of the British Isles. I knew Group W was where I wanted to work.

Dr Deacon, it turned out, had a great respect for mathematicians. Himself a chemist, who had made his name by arduous and painstaking hydrographic observations in the Southern Ocean, he nevertheless appreciated the vital contributions that Fritz Ursell, in particular, had made to the design of the wheel analyser and the basic wave theory behind it. Based partly on the Cauchy-Poisson theory of waves from a localised source, and partly on a war-time Ministry of Supply report by Harold Jeffreys, the theory depended on the fact that the energy of ocean swell travels away from a storm not with the speed of the wave crests but with the group-velocity, which in deep water is just half that speed.

So I was conscious, on my first day of work, that my mathematical training would be put to good use, and any theoretical contribution, if it was significant, would be appreciated.

Dr Deacon had all the main scientists and engineers together in that one room, without partitions, so that they could easily communicate. Fritz Ursell had the desk next to his. I was placed nearly at the other end, close to Norman Barber, who was a physicist, and from whom I learnt the value of a 'physical' approach to mathematical problems. He set me to work first on his new idea for an 'electrode flow-meter'. Some South African physicists had suggested measuring flow near the seabed by detecting the emfs (electro-motive forces) induced by passage of the water across the Earth's magnetic field (using Faraday's law). Barber proposed two improvements: (1) to use an alternating magnetic field, so that the 50 Hz component would distinguish the signal from the ever-present geomagnetic noise; (2) to raise the instrument a foot or two above the bottom, so increasing the effect of the water flow and perhaps doubling the sensitivity. Barber's suggestion to me was to model the magnetic field, which in reality would be produced by a circular coil with vertical axis, as a magnetised ellipsoid, and hence find where best to place the pairs of electrodes.

I remember, on my first day, scratching my head over Faraday's law in orthogonal ellipsoidal coordinates and at lunch-time being mortified that I had not found the solution. At Cambridge I had been used to taking at most one hour over a question in the Maths Tripos. But fortunately by the end of the day I had the answer, and to cap it, a beautiful general theorem: if a conducting fluid moves irrotationally in a magnetic field, any maximum or minimum in the induced electrical field has to lie on the boundary. This meant that the best place for each electrode was always on the solid boundary – namely on the ellipsoid itself. This idea was incorporated by

Frank Pierce into the initial design. The first instruments were about 18 inches in diameter. Later on, much smaller models with a diameter of a few inches were routinely used for measuring a ship's speed through the water, and were adopted by other ships and oceanographic institutions.

This study led to much broader consequences. After the war two Post Office engineers, D.W. Cherry and A.T. Stovold, were restoring the broken telephone communications between England and France. In a letter to *Nature*[17] they described how in testing the cross-Channel cables they had earthed the copper core at one end, while at the other they measured the voltage between the core and its metal sheath. They found a voltage of about ±0.4 V, fluctuating with tidal period (12 hours 24 minutes) and suggested it was due to tidal streams in the English Channel moving across the Earth's magnetic field. Norman Barber, seeing my success with elliptical coordinates, suggested that I model the English Channel as a long, straight channel of semi-elliptical cross-section, embedded in a conducting medium. This I did, and discovered that the induced voltage would be partly short-circuited by the seabed. The resulting voltage would depend critically on the depth of the channel and the electrical conductivity of the seabed compared to that of sea water. So we could get an estimate of the seabed conductivity, perhaps down to depths comparable to the width of the channel. Remarkably, this also led to the measurement of the wind-stress on the water, as will be described later.[18]

Meanwhile I was taking a close interest in ocean waves. Barber and Ursell had considered only two one-dimensional aspects: distance and time. In reality sea waves are often choppy and short-crested. I wrote a brief report showing how one could extend the idea of a frequency spectrum of water waves to two spatial dimensions, and one of time. Norman Barber took this up and conceived the idea of Fourier-analysing the sea surface by treating a transparent aerial photograph as a diffraction grating, using a light source instead of X-rays. The idea worked but the results were difficult to interpret quantitatively. He also suggested using the tilting motions of a floating buoy, and its vertical acceleration, to gain some information about the spread in direction of the wave energy. These ideas were collected together in a 1946 ARL report to which Tom Tucker also contributed: 'Four theoretical Notes on the Estimation of Sea Conditions'. It attracted attention both in England and among oceanographers in America.

As mentioned earlier by Jack Darbyshire, George Deacon was also interested in microseisms. These are small oscillations of the ground, present on many seismograms. Like sea waves they tend to occur sporadically, in so-called 'microseism storms'. The first studies seem to have been made at the Vatican Observatory in Rome in the early 1900s. In the 1930s French investigators had found a connection between the microseisms recorded in Paris and Strasbourg and the ocean swell recorded on the coast of Morocco. The microseisms were somehow related to distant storms south of Iceland, but with a certain time delay – the further the storm from North Africa the greater the delay of the swell. What could be the cause of the microseisms, and could they be of any use in predicting ocean swell? Perhaps they would even indicate the weather in under-observed parts of the ocean (in those days there were no orbiting satellites).

At Teddington we were close to the famous weather observatory at Kew, where there was also a recording seismograph. Deacon compared the period and amplitude of the microseisms at Kew with the period and amplitude of the swell at our recording station at Perranporth in Cornwall. There was a clear correlation. Surprisingly, however, the period (time interval between successive crests) of the microseisms was just half that of the corresponding sea waves. What could be the reason?

Just then we were visited by a French engineer from NEYRPIC, an organisation for hydraulic research based in Grenoble, in the south of France. He was F. Biesel. He pointed out to us a long paper, published in 1944-45 by M. Miche, and concerned with the theory of water waves. Among other things Miche had studied standing water waves (like those sloshing from side to side in a bath) and had carried his calculations to a second approximation, that is to say he included terms proportional to the square or product of the particle displacements. Significantly and surprisingly Miche had found, at the second approximation, a term in the expression for the pressure, which was not dependent on the vertical coordinate, that is on the depth of a particle below the surface. This term fluctuated in time with just twice the fundamental frequency of the standing wave. Could this be a clue to Deacon's observations?

Miche's mathematical analysis was long and complicated. I hit on a much simpler way of doing the calculation, and also a physical reason for the doubled frequency: a standing wave was just like a swinging pendulum, with the bob going from side to side. Twice per cycle the bob was up (higher) and twice it was down (lower). This caused a vertical reaction on the hinge, at twice the basic frequency. For water waves the corresponding motion was similar and easy to compute. It produced a double-frequency pressure, independent of depth. It was as if someone were jumping up and down on a floating raft, at twice the wave frequency.

Moreover, this happened not just for perfect standing waves; any pair of opposite waves of the same frequency and different amplitudes would also have the same effect. The double-frequency pressure term was just proportional to the product of the two wave amplitudes. Furthermore the calculation was equally simple for waves in more than one dimension, which I had already thought about.

Fritz Ursell became interested in my conclusions and supplied his own interpretation, based on the idea that the centre of gravity of the whole wave train was raised and lowered twice in a cycle; the only external force that could cause this would be a pressure variation deep down. Together we wrote a short theoretical note to *Nature*.[19] With George Deacon we also visited the Geography Department in Cambridge and attended a symposium at which we (or rather Fritz) explained our ideas.

George Deacon used to say to his staff, "Don't work too hard!" which was well meant but in fact made them work all the harder. During the spring and summer of 1946 I needed a rest, and so Norman Barber suggested that I spend a month looking after some electrical equipment at Cawsand Bay, near Plymouth, where during the war there had been a degaussing range. Closed loops of electrical cable were laid on the seabed, to measure the effectiveness of degaussing the ships as they passed over the loops. The cables were connected to highly-sensitive galvanometers installed at an underground storehouse named Pier Cellars. Before the war it was used by local fishermen for preserving their catch. There was an unoccupied cottage on the hill above, and a small lighthouse down by the entrance to the harbour. Two of the degaussing loops had been reconnected, so that now they were expected to register the motion of the tidal streams, as explained above.

All went as planned, except that in order to be closer to the recording galvanometers, which were tricky, and to change the paper at night, I had my bedding and cooking stove (all borrowed from Devonport Dockyard) moved, with the consent of the Admiral, from the cottage above down a steep flight of steps to the lighthouse, by means of ropes, planks and pulleys. In that lighthouse I spent many restful hours listening to the sound of waves coming against the harbour wall.

Back in Teddington my first concern, after analysing the records from Pier Cellars, was to develop a quantitative theory for estimating the microseism amplitude, that is the vertical displacement of the seabed, caused by a given sea state at a given distance from the storm. Some of the mathematical tools needed, for example the idea of a continuous, two-dimensional wave spectrum, were not then available and I had to invent them. One unexpected factor was the compressibility of sea water. At certain depths, equal to about 1/4, 3/4, 5/4 times the length of a compression wave having the same frequency as the applied pressure oscillations, the resulting displacements could be amplified by a factor of order 5. This 'organ-pipe' effect appears from recent observations to be important.[20] In North America, for example, the largest microseisms come from the Labrador Sea, where the depth is about 3,000 m. Similarly in Northern Europe and Scandinavia the largest microseisms come from the deep water off the edge of the continental shelf.

In 1947, however, all the theory had still to be worked out. I spent much of my time in the library of the National Physical Laboratory next door, where there was a good selection of scientific journals; others could be obtained on loan from other libraries. Back at my desk in Group W I worked with a small Brunsvega calculator, postwar loot from Germany. Finally in 1948 my results were put into an Admiralty report, but before submitting it for publication I waited until I had done further calculations and some wave experiments at Cambridge.

One last project at Teddington was again at the suggestion of Norman Barber. He pointed out that according to my model of the English Channel as a long straight channel of (semi) elliptical cross-section, the tidally-induced electrical potential would have a maximum at the shoreline on, say, the English side, and a minimum at the shoreline opposite. But the potential at infinity, that is to say at a great distance inland, would be zero. Thus there would be a gradient of electrical potential in the land, at right-angles to the shoreline. If one were to embed two electrodes in the ground on the same side of the channel, but at different distances from the coast, one ought to detect a difference in voltage between them. This would be due to an electrical current being pumped into the land by the tidal dynamo. It should fluctuate with the tidal streams.

To carry out a test Barber proposed using an abandoned gunnery range near Warbarrow, in Dorset. Here were some abandoned cottages, with water supplied from local wells. By lowering silver electrodes into the wells, one could get good electrical contact with the ground at that point.

An expedition was arranged. Collins built a beautifully neat recording galvanometer, contained in a flat box. I travelled down to Dorset by road in the Group W van, driven by Rick Hubbard, an assistant. By arrangement with the Army no guns were fired that month. Two pairs of electrodes were placed in the wells of cottages in a north-south line at distances of 2 to 4 miles. I stayed in a guesthouse nearby and went over from Wool railway station by bicycle. To my delight the records, when developed, showed a distinct tidal fluctuation, of the order of 100 mV, which agreed in phase and amplitude with the predicted tidal streams in that part of the English Channel. So we had measured the tidal streams without getting our feet wet!

Later-on this same method was used to measure fluctuations in the strength of the Gulf Stream off the coast of Florida; also the flow in Cook Strait, New Zealand, between North Island and South Island; in the Menai Straits between North Wales and Anglesey, and in other places. Most notably, however, K.K. Bowden at NIO compared the cross-channel voltages between England and France with the predicted tidal streams and was able to infer the additional flow caused by the wind. This provided a novel means of measuring wind-stress at the sea surface.

In September 1948, having completed my three years' 'work of national importance', I returned to Cambridge to study for a PhD. However, it seemed to me that I hardly left Group W. Under Harold Jeffreys and then under Robert Stoneley (of 'Stoneley waves') I was allowed to pursue the same investigations as before, reporting to my research supervisor only once at the end of each term.

As to microseisms, my first inclination was to verify an approximation I had made, whereby the water in the upper layer of the ocean (about half a wavelength thick) was treated as incompressible even while generating compression waves below. The approximation turned out to be valid. Then I designed and carried out some experiments to test the theory. Being allowed the use of a 30 ft-long wave channel in the Geography School, I set out to adapt an old navy hydrophone so as to measure the pressure fluctuations at different depths below standing or partly reflected water waves. In this I had the help of R.I.B. 'Rib' Cooper in the Department of Geodesy and Geophysics at Madingley Rise. Near the surface, I found a mixture of first and second harmonics, but deeper down, at a depth of half a wavelength or more, only the second harmonic, or doubled-frequency, pressure fluctuations were present, and these persisted to greater depths without attenuation, just as predicted.[21] When I demonstrated this phenomenon to George Deacon and his friend Vaughan Lewis in the Geography Department, they were astonished and delighted.

Additionally, by simply measuring the amplitude of the pressure oscillations, which were proportional to the product of the amplitudes of the incident and reflected wave trains, we had to hand a new and simple tool for determining the amount of energy reflected from different kinds of beaches and submerged obstacles.

My second main investigation at Cambridge also involved water waves. This had to do with the movement of sand by waves in shallow water. In 1944, R.A. Bagnold, founder of the Long Range Desert Group during the North African campaign and author of a classic book, *The Physics of Blown Sand*, had done experiments at Imperial College showing that a thin layer of water near the seabed, besides oscillating to and fro with the waves, had a remarkable mean drift towards the beach. (That was outside the zone where the waves break; inside, the flow is reversed.) How to account for this remarkable forward drift?

By applying a boundary-layer theory, where viscous forces are fully taken into account, I found that the forward velocity could be quite accurately predicted; there was good agreement between my theory and Bagnold's measurements. This of course also delighted Vaughan Lewis and others in the Geography Department who were interested in how beaches are built up or destroyed by sea waves.

In the summer of 1950, I joined Deacon and Lewis in their joint expedition to Chesil Beach, west of Portland Bill, to study shingle movements due to waves. It was my first meeting with Margaret Brenda Deacon, then aged eight.

In 1951, the year I became a PhD, I was granted a Commonwealth Fund Fellowship to study oceanography in the USA both at Woods Hole, with Henry Stommel, and at La Jolla, with Walter Munk. During this very productive year I kept in close touch with George Deacon. The year ended with our both attending a conference on microseisms at Harriman, in upstate New York. All the leading workers on microseisms were present, including Frank Press and Maurice Ewing, from the Lamont Geological Observatory. I gave an outline of my theory, followed by comments from Deacon and others.[22] It was well received.

In a given sea state, not all waves are the same height. At about that time, there was considerable discussion among ocean engineers as to what is the statistical distribution of wave heights or amplitudes? While working in La Jolla, I produced a simplified theory based on the assumptions of a narrow-band energy spectrum. From this I calculated the values of some commonly used parameters, such as the 'significant wave height' (i.e. the mean height of the 1/3-highest waves); the highest wave among 60 or 100 consecutive waves, and so on.[23] The theoretical ratios between these quantities turned out to be in good agreement with observation. Since then the theory has been much used and developed.

With Henry Stommel and Melvin Stern at Woods Hole I also wrote a theoretical paper on the use of 'towed electrodes', i.e. pairs of electrodes towed behind a ship, for determining the speed of local currents; and on the conditions necessary for the method to work.

So it was not surprising when after two years back in Cambridge with a junior research fellowship, I decided to accept George Deacon's invitation to rejoin my old colleagues in Group W and other more recent recruits such as Henry Charnock, David Cartwright and Jim Crease, in the new National Institute of Oceanography at Wormley. It seemed to me that I had hardly been away.

Several of the original members of Group W had returned to civilian life. Norman Barber had briefly moved to the Admiralty's headquarters in London "to learn about administration", as he said, but found it mainly involved "the circulation of dockets". Following the sad death of his wife he emigrated to New Zealand. Clifford Mortimer had joined the laboratory of the Freshwater Biological Association on Lake Windermere in Cumbria and became director of the Millport Laboratory of the Scottish Marine Biological Association before being appointed Director of the Great Lakes Research Centre in Milwaukee, Wisconsin. Fritz Ursell was appointed a Lecturer in Mathematics at Manchester University, and afterwards in the new Department of Applied Mathematics at Cambridge. All three subsequently became distinguished University professors. But all of us, together with those who stayed behind, owed to George Deacon a great debt: the unfettered opportunity to do creative work in the expanding and exciting field of ocean science.

Life in the Oceans

The pre-war Discovery Investigations had conducted detailed studies of the biology and hydrography of the Southern Ocean aimed primarily at providing a scientific foundation for the whaling industry. Peter Foxton, a biologist who sailed on the first postwar cruise of RRS Discovery II *and Martin Angel, later head of the IOS biology group, describe how this early work developed into a broader understanding of open ocean ecosystem structures and dynamics that today provides much information against which current problems and changes in the oceans can be assessed. Howard Roe, who became Director of the Southampton Oceanography Centre, describes the postwar research into whales and the role of NIO scientists in the development of the International Whaling Commission and subsequent conservation of whale populations.*

5

Ocean ecology

Peter Foxton and Martin Angel

The legacy of the Discovery Investigations

Early in the discussions that led to the creation of the NIO, it was decided that the Discovery Investigations, with their long experience of research in the Southern Ocean, would be incorporated within the new institute. So it was that the Discovery Investigations' Director, Dr Neil Mackintosh, together with James Marr, T. John Hart, Henry Herdman, Helene Bargmann and a new recruit Robert Clarke, joined the new NIO on 1 April 1949. This transfer of a well-established organisation with a tradition of biological research and a nucleus of experienced seagoers laid the foundations for future marine ecological research at NIO.

Plans to restart the Discovery Investigations Antarctic research programme, interrupted by WWII, were at an advanced stage. Both research vessels, RRS *Discovery II* and the RRS *William Scoresby* had seen war service, and were being refitted and re-equipped. Additional scientific staff were needed for both ships offering a marvellous opportunity for graduates at a time when there were few vacancies in marine science. Urged by his Professor, Peter Foxton sent off an application and was delighted to be called for interview in London. Dr Deacon, with characteristic thoughtfulness and concern, concluded the interview by emphasising the hardships and discomfort of working at sea and warned that they could be away from the UK for up to two years. Peter was relieved to receive an offer of a temporary appointment as a biologist on *Discovery II*, which he hastened to accept, so avoiding plan B – becoming a teacher. Having been permitted to defer National Service, he reported for duty on 5 October 1949, joining the other new biologist, Peter David, and also Ronald Currie, who was to be the chemist on the *William Scoresby*. Later recruits were Roland Cox, a chemist, and three scientific assistants, Roy Plummer, John Hooper and Ed Childs. Their leader was the vastly experienced and resourceful Dr Henry Herdman.

With no central headquarters the new staff were dispersed, the chemists to the Marine Biological Association at Plymouth, and the biologists to the Discovery Hut, home of the Discovery Collections and located at the back of the Natural History Museum in London. The curator was the kind, elegant Dr Helene Bargmann who arrived each morning accompanied by her Pekinese dog. The Hut was a basic wooden structure designed for storage rather than research but there were benches, microscopes, sinks and an unlimited supply of formalin, all that was needed for a crash course on Southern Ocean plankton. Our mentor was the redoubtable James Marr who as a boy scout had sailed on Shackleton's last Antarctic voyage on the *Quest*, had been a member of the British Australian (and) New Zealand Antarctic Research Expedition (BANZARE) campaign, and had extensive pre-war seagoing service with the Discovery Investigations. During the war he commanded Operation Tabarin which established bases in the Falkland Islands that were later used by the Falkland Islands Dependency Survey. Returning from the Navy he resumed his seminal study of krill, *Euphausia superba*, the primary food of the Antarctic baleen whales. Under his guidance we learned how to identify plankton, how to preserve and label samples, and how to identify adult, juvenile and larval stages of krill. Behind locked doors the research departments of the Natural History Museum were fascinating places staffed by some of the leading taxonomists of the day, several of whom would drop in for the ritual of Dr Bargmann's morning coffee and the Times crossword, and also advise us novices.

Dr Mackintosh was a frequent visitor from his office in Whitehall. During his rather formal visits he would monitor progress, outline his plans for the forthcoming cruise (also embodied in a series of detailed memoranda), allocate us our specific tasks and invariably take notes. The extensive work of the pre-war commissions had provided a broad basis for describing the oceanographic conditions of the Southern Ocean and its general ecology. But the geographical coverage was uneven, being concentrated mainly in the Atlantic sector, with relatively few wintertime observations anywhere. The plan was to fill in the gaps by making observations in the regions and times of year where data were scant or non-existent. The biologists would make routine collections using the sampling methods and protocols used on all previous cruises to ensure comparability.[1] The samples were to be roughly sorted and analysed at sea, with special attention being paid to the occurrence of krill. Previous research on the zooplankton in the SW Atlantic sector had established the existence of faunistic zones. Did these occur in other sectors? To address this it was proposed to measure the biomass of the zooplankton caught with a standard net, the N70V (70 denoting the mouth diameter in cms and V a vertical haul), to compare spatial and temporal variations.

Each biologist was encouraged to focus on a particular species and to study its distribution and life history, thereby continuing the well-proven Discovery Investigations' tradition for scientists to develop specific expertise in systematics and pursue their own lines of research within the framework of the overall scientific programme. To new recruits the plans appeared highly ambitious, but with two years ahead of us anything seemed possible.

The refitted *William Scoresby* sailed from London for Capetown on 11 January 1950 under the command of the experienced Lt Cdr A.F. Macfie and with a scientific team of John Hart (who left in Cape Town), Robert Clarke and Ron Currie. The plan was to first conduct a major survey of the Benguela Current upwelling region, then to engage in experimental fishing with a variety of nets and traps (the crew included three fishermen), and finally to conduct a major programme to mark whales during their winter migration to tropical waters off the northwest coast of Australia and South Africa.

The outward passage was dogged by mechanical and electrical problems, and sickness amongst the ship's company which reduced the time available for science. Nevertheless a successful survey of the hydrology and ecology of the Benguela current off southwest Africa was conducted in March, confirming the presence of upwelling near the coast which produced an area of exceptionally high productivity. (Following a change to the subsequent programme it was possible to resurvey the region between September and October when upwelling was absent.) The ship was frequently awash and on one occasion the doors of the chemistry laboratory were stove-in. Ron Currie made many of his analyses up to his knees in water but, despite this, the comprehensive data set that resulted provided the basis for a Discovery Report by John Hart and Ron Currie that became a classic account of an upwelling region.[1] (Ron Currie was later to join *Discovery II* in March 1951 as chemist when Roland Cox returned home.)

From Capetown the ship proceeded into the Indian Ocean to undertake fishing experiments off Mozambique and in the Agulhas Current, including an abortive attempt to catch a coelacanth using a large trap and drop lines. On 27 June the ship sailed from Mauritius for Fremantle aiming to arrive in time for the humpback whaling season. Strong headwinds and resultant shortage of fuel led to the crossing being abandoned, and the ship returned to Mauritius. A revised programme followed, included more whale marking off South Africa and the additional survey of the Benguela Current. On 18 November the voyage ended at Plymouth. The *William Scoresby* was laid up and eventually sold for scrap for £1,900, an inglorious and undignified end for a small vessel that had a distinguished tradition of facilitating the highest quality research, often in the most difficult conditions.

The refit of *Discovery II* was intended to restore her to her pre-war state and to improve the accommodation: the location of winches, deck equipment, and layout of the laboratories were exactly as they had been on her last 1937-39 commission. Technical developments resulting from the war included the installation of radar, which necessitated the addition of a radar technician to the ship's complement, and Kelvin Hughes echosounders. Cdr John F. Blackburn DSO was appointed as Captain and the officers and crew were mainly recently demobbed from the Royal Navy. Cdr Blackburn was to prove a formidable skipper, who inspired absolute confidence. Somewhat brusque and a man of few words he was always happiest at sea, unless international cricket was on offer! We were allocated standard Navy clothing: duffel coat, white sweater and short rubber boots, with a leather coat and boots for working on the exposed foredeck. The refit suffered so many delays that there was no time for a shakedown cruise to familiarise the officers, crew and the scientific team with the workings of the vessel. So when the ship eventually sailed on 20 April 1950, the outward passage to Australia effectively became a prolonged shakedown. It was a salutatory learning period for all involved and many defects were soon revealed. Two days into the cruise, the ship returned to Plymouth for essential repairs and was delayed there until 10 May.

Dr Mackintosh joined for the passage to Malta and instructed the biologists in using the biological sampling gear. The chemists were taught how to collect water samples by Henry Herdman, shallow ones from the mid-ship winch and deep ones from the forward fo'c's'le winch. At each sampling station we worked a standard routine. The ship hove to usually at 20.00 h. Phytoplankton was sampled with the fine mesh N50V hauled from 100 m to the surface. Zooplankton samples from known depth intervals were then collected down to a depth of 1,500 m by a sequence of hauls with the N70V net. At the same time water samples were collected, first from shallow depths using the well deck winch, and then the deeper sample on the fo'c's'le winch. All vertical sampling was conducted on the port side and success depended on the skill of the deck officers in manœuvring the ship to maintain the wires vertical both fore and aft. When heaving-to, the ship's bow was positioned with the wind just to starboard, the aim being to balance the speed at which the wind was blowing the ship backwards with forward thrust from the ship's propeller. This was hard enough when winds were light and steady and sea and swell were calm, but when the wind was gusting with cross swells and strong subsurface currents the task became extremely difficult and on occasion impossible. Utmost skill and constant attention was required of the watch-keeping officer on the bridge, bearing in mind the slow response of the ship to its steam driven, low-revving, single screw. Once the vertical work was completed other nets were towed both horizontally and obliquely using the main aft trawling winch. On average a station would take four hours to complete, but this was much longer in adverse conditions. By the time we berthed at Malta on 21 May we had worked seven stations along a transect that crossed the continental shelf from shallow to deep water to the west of the English Channel, and several stations from the Bay of Biscay to the Western Mediterranean.

The cruise proceeded via the Suez Canal and the Red Sea into the Indian Ocean. After a brief stop at Colombo we worked a line of stations from the equator to 30°S along 90°E in a region that still remains poorly explored. Passage through the Indian Ocean proved unexpectedly eventful. One lunchtime in June the familiar beat of the engine changed alarmingly, heralding a major failure of an engine bearing. For some 24 hours we drifted while repairs were affected. Fortunately the weather was calm but prudently a sail was rigged from the foremast, giving us a speed of 0.5 to 1 kt. This was the only time that the sail was ever used and it was reassuring that the engineers were able to cope with a major repair while at sea. Confidence in the engineering staff was again tested when a few days later the ship suddenly veered off course and proceeded to steam in circles. The steering engine had broken down and repairs were impossible at sea. Thereafter two seamen were stationed in the steering flat to manually helm the ship using the emergency wheel.

November 1950, Discovery II *in the Antarctic, the scientific team dressed for action. (Left to right.) P. David, J. Hooper, E. Childs, R. Plummer, P. Foxton, R. Cox.*

Lack of maneuverability meant that the scientific programme had to be abandoned and the ship limped into Fremantle on 6 July.

Delays due to repairs meant that we had to modify our planned sampling programme, and we left Fremantle for the Antarctic on 16 August. This leg was a rude introduction to the rigours of the months ahead. Gale succeeded gale with constant and monotonous regularity, with an occasional storm thrown in for good measure; only on three days did the wind drop below Force 6. Station keeping was always extremely difficult and frequently sampling was abandoned while the ship rode out the storm. Nevertheless

Left: *Roland Cox on the forward hydrographic winch setting a reversing water bottle.*
Right: *Ed Childs and John Hooper taking a sample from the Nansen-Pettersen insulated water bottle.*

most of the planned stations were worked, though with a heavy toll on gear and damage to the ship. Having to steam into the wind for long periods also resulted in a fuel shortage, and finally the Captain took advantage of favourable winds and sailed for Melbourne rather than Fremantle.

Having experienced and survived our first Southern Ocean cruise, the crew and scientists settled down to a life of long periods at sea interspersed with brief spells in port for refuelling, re-victualling, recreation and receiving news and mail from home. Port calls were followed all too soon by farewells and departure for another Antarctic cruise. Life on board was organised along Royal Navy lines. There were separate messes for Petty Officers, Crew, Greasers and Boys. The scientists shared the wardroom with the Captain and Officers. Dress for meals was formal (jackets and ties) and we dined off white linen tablecloths with personal napkins, attended by two stewards. When at sea, fiddles were fitted to the tables to prevent plates and cutlery sliding onto our laps and frequently the tablecloth was dampened as an extra measure. Meals for a complement of 56 were prepared and cooked in a small hell-hole of a galley often under extreme conditions. Cooking during rough weather often resulted in loud crashes followed by volleys of expletives from the galley. Such catastrophic events resulted in radical changes to the published menus with sandwiches becoming the order of the day. Facilities for storing fresh food were limited, so tinned food predominated after a few days and the menus took on a monotonous predictability; 'herrings in' and the memorable pink bologna sausage, interspersed with 'yellow peril' (smoked haddock) and 'babies heads' (steak and kidney pudding). The Captain sat at the head of the table with the Chief Scientist on his right and the first mate on his left; at breakfast an out-of-date Times newspaper would appear that eventually found its way to the wardroom. For entertainment the wardroom boasted an upright piano and a cabinet windup gramophone of uncertain vintage and a limited collection of 78 records. The sound of Beniamino Gigli singing an aria from *La Bohème* as the ship lay motionless in the dead calm water at the edge of the pack ice was something to be experienced. In lieu of a bar there was a drinks cupboard (locked) which the junior scientists were discouraged from using when at sea. Following Navy tradition on Saturday, following Captain's rounds, each crew member, boys excepted, was issued with a tot (¼ pint) of Navy Rum (100% proof alcohol!). The man to sight the first whale and the ice edge earned an extra tot. (The dining formality, food, and rum tots persisted until the late 1960s on board *Discovery*; the gramophone did not!)

Each cruise consisted of long lines of our standard stations. At a few stations a variety of other nets were used and, since little was known about the topography of the seabed, echosounder surveys were routinely conducted while underway. Daily meteorological observations were made and transmitted by the radio operator, who was our only contact with the outside world. After each station there was much to occupy the scientific team. Water samples had to be analysed, usually within 24 hours, biological samples had to be preserved, labelled and roughly sorted. The displacement volumes of the N70V samples were routinely measured, and all zooplankton samples were examined for krill and other euphausiid species. I (PF) developed an interest in pelagic tunicates or salps, gelatinous animals often present in enormous numbers, that were little known at the time. These filter-feeding animals, together with other 'jellies', are now known to play important roles in oceanic food chains. Peter David started one of his main research interests, a study of the general ecology of chaetognaths (arrow worms). These carnivorous predators are another important group of gelatinous plankton that often control the abundances of other species. Peter, already an accomplished photographer, saw the potential for compiling a permanent record of living plankton with his Leica and Rolleiflex cameras and the enlarger installed in the ship's small darkroom. Suitable colour film was not available so these early pictures were taken in black and white. So began a long-term

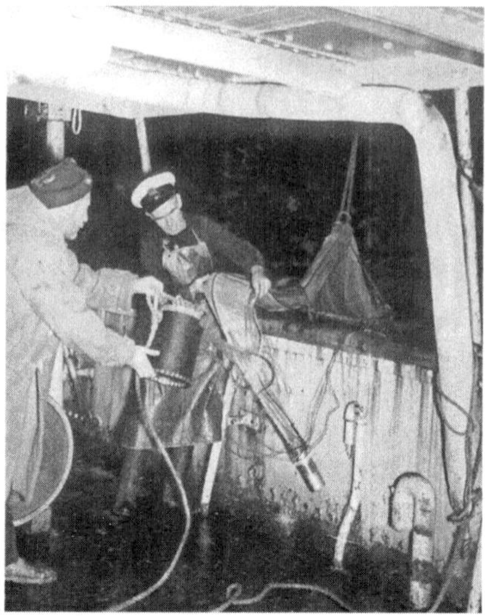

Left: *Peter David keeping the log, aft rough lab.*
Right: *Washing down the N70V catch. Harry Moreton (bosun) in the uniform cap.*

interest in assembling a library of unique images of live plankton and larger nekton: pictures that continue to be used in publications today.[2]

If these activities were not enough, the inventory of scientific equipment on board was a veritable naturalist's Pandora's Box, providing for nearly every eventuality. To compile a photographic record of the cruise there were two Sanderson wooden plate cameras which provided a challenge to Roy Plummer. The armoury was kept securely by the Captain, and comprised two Holland & Holland shotguns, a box of whale marks, and a small harpoon gun. There was entomological equipment for preserving and mounting any insects that might come aboard, and at the suggestion of that doyen of biological oceanographers, Professor Alister Hardy (himself one of the original *Discovery* scientists), we attempted to sample 'aerial plankton' with a small net flown from the aft mast. However, unless we were close to land the net sampled only soot from the funnel. Even so we assembled a small collection of insects which was deposited at the Natural History Museum in London. We had a plant press and supply of blotting paper for preserving terrestrial plants. Whale and bird observations were routinely recorded, the latter by the keen 3rd Officer, George Selby-Smith. We also had an oological outfit for blowing birds eggs, and a large tin of arsenical soap for preserving the skins of birds: though the only preserved bird was eaten by the ship's cat, which remarkably survived! There was little opportunity to use most of this equipment but we did attempt to record the structure and colour of any notable plankton species using the water colour paints and sketch pads that we found in one of the drawers, though our efforts never matched the exquisite sketches of Alister Hardy.

Cartoon by Dick Burt, the Netman, of the crew taking on fuel from a cache of oil drums left for us at Kerguelen during the long winter circumpolar cruise.

There were inevitable delays to the programme, which was frequently modified to cope with adverse weather, particularly during the arduous winter circumpolar cruise of May to September 1951. This involved calls at the Falkland Islands, Kerguelen, and Heard Island before finally berthing at Fremantle. In all we spent nearly two years working in the Southern Ocean, steamed some 39,000 miles there and achieved all of our main objectives. On the homeward passage we repeated the line of stations in the Indian Ocean along 90°E from 30°S to the Equator; diverted briefly to the Maldives in an unsuccessful search for a species of petrel; worked stations in the Red Sea and Mediterranean, and finally arrived in Plymouth on 6 December 1952. This was the last cruise by *Discovery II* to the Southern Ocean and one of the last voyages entailing such an extended period away from the UK. It was the end of an era.[3]

In retrospect the cruise proved to be a life-changing experience for the young graduates who had set out from Plymouth in 1950. We all returned wiser and with few illusions about the practicalities of shipborne research. We learned that programmes had to be well planned with comprehensive sampling in time and space followed by painstaking lengthy analysis. Most importantly we came to realise that successful work at sea depended on a collaborative team effort. Having savoured the excitement and rewards of oceanographic research, David, Foxton, Cox and Currie all continued their careers with NIO. Three stalwarts in the crew, Harry Moreton (Bosun), Charlie Fry (Donkeyman) and Dick Burt (Netman), continued to serve on the research ships for many years, the latter making important innovative contributions to gear design and handling.

At home some of us returned to the Discovery hut, where we were joined by newcomers Arthur de C. Baker and Keith Andrews. Meanwhile Dr Mackintosh had been considering the future of work in the Southern Ocean, which he passionately believed should be continued. He proposed that the main object of future research should be

> *to draw up a comprehensive statement of conditions in the area covered, in a form which will (a) add to what is known of the large scale physical, chemical and biological processes which go on in the oceans as a whole, and (b) constitute a fund of information on the Southern Ocean which in particular is calculated to be of direct value both to other research in the region and to problems of immediate practical and economic importance.*

Continued studies of the life cycles of selected species would be important to provide comparison with Marr's work on the ecology of krill, and, perceptively, Macintosh suggested that the research would eventually merge with the new work of NIO.

New Opportunities at Wormley

In 1953 the move to Wormley started, bringing the various research and support groups together on the one site. It was a defining event for the biological group and for the future of its research. Wormley offered not only a multidisciplinary environment but also access to specialist electronic and engineering expertise and support. This created a unique research environment rich in novel opportunities and technical innovation that continues to this day. An exciting and stimulating era ensued in which we all benefited from the enthusiastic encouragement of the ever-approachable Dr Deacon.

Priority continued to be given to ongoing and new research on the Discovery Collections, resulting in the publication of a substantial series of reports on, for example, chaetognaths (David), euphausiids (Baker), Antarctic krill (Marr), salps (Foxton), specific studies on important species of zooplankton, and on the Benguela Current upwelling system.[1] These and many other studies established the general and specific pelagic biology of the Southern Ocean and the relationship of this to the physical oceanography; as such they have been essential for much of the subsequent biological oceanography in the Antarctic.[1] As Southern Ocean projects were completed, opportunities arose to explore new problems resulting from *Discovery II*'s operations in the North Atlantic. Arthur Fisher joined NIO at about this time and after admitting to an enthusiasm for butterflies, found himself in the Biology Group. He was quickly introduced to work at sea occupying a key role as assistant to Ron Currie.

One of the first new research projects initiated by Dr Mackintosh came from his realisation that Sperm whales and their ecology offered great potential for study. At the time they were hunted commercially in the Azores and Madeira, where Robert Clarke was already working. The abundant remains of squid mouthparts (beaks) in the stomachs of Sperm whales suggested that oceanic cephalopods were a major part of their diets. Knowledge of the ecology of squid populations could thus be a key to understanding the ecology of the whales. But squid are adept at avoiding nets, and are rarely caught by conventional means: so how were they to be studied? Our first attempts in 1952 using baited hooks attached at intervals to the 4 mm plankton wire failed dismally, in marked contrast to the ease with which they could be caught with fishing rods and hand nets at the surface at night! Then in 1956 we tried using the deep-sea camera that was developed by Tony Laughton to photograph the seabed. A simple modification of attaching the bait to a trigger meant that a photograph resulted if the bait was taken. To our delight we obtained a unique photographic record of the squid *Sthenoteuthis (Ommastrephes) pteropus* taking baits at various depths.[4]

Following this modest success the programme on oceanic squid took on a new momentum with the appointment in 1959 of Malcolm Clarke. He initiated a study of the extensive samples of squid beaks routinely

Photographs of a squid taking bait, getting hooked and then attacked by a second squid (R). Their intertwined tentacles can be seen. The camera was triggered when the squid pulled on the bait.

collected from Sperm whale stomachs. In this he was assisted by Neil MacLeod, and by painstaking measurements of hundreds of beaks he was able to match these with ones found in whole or partially digested squid which sometimes also turned up in the stomachs. The results dramatically increased our knowledge of both oceanic squid and sperm whales.[5]

The early postwar biological cruises of *Discovery II* in the North Atlantic aimed to investigate the ecology of the pelagic ecosystems in the seas around the Azores, Madeira and the Canaries. Our interest lay in the vertical structure of the communities in relation to diurnal migration, (the upward migration towards the surface carried out by most oceanic animals at night), and the effects of this on the deep scattering layers recorded by echo sounders. Initially the sampling gear and methodologies were the same as those used by Discovery Investigations; the rationale being that standardisation allowed semi-quantitative interpretation of the data in the absence of absolute measures of performance. The main tool then for sampling zooplankton was the N70V net, which could be closed using a throttling technique. Trials were carried out that demonstrated the effectiveness of this closing method and thereby confirmed the reliability of the earlier Antarctic samples.[6]

This early critical assessment of plankton sampling stimulated much further evaluation and development of sampling techniques. Ron Currie was beginning to use the ^{14}Carbon method devised by Steeman Nielsen for measuring primary production, using large water bottles for sampling phytoplankton to quantify organic production of oceanic waters. But there was no way of measuring the efficiency or amount of water filtered by the plankton nets, and no effective ways of catching the larger macroplankton and micronekton (typically shrimps and fish) thought to make up the deep scattering layers. To address the plankton problem a new net was developed that closed like the N70V but which incorporated a flowmeter to measure the amount of water being filtered by the net and which also recorded the depth of the net. Dicky Dobson, a clever, albeit eccentric engineer, designed the flowmeter and had a great facility for translating the vague ideas of scientists into functional designs from sketches roughed out on the back of old cigarette packets. This new net proved to be both efficient and quantitative, i.e. the numbers of plankton caught could be related to the amount of water filtered.[7]

Having improved the method for sampling zooplankton, our attention turned to improving sampling of the larger animals. Large conical nets of mouth diameter 1 m, 2 m and even up to 4.5 m had been used on the Discovery Investigations but they had limitations and alternatives were needed. In 1956 trials were carried out using a net designed at Scripps Institution of Oceanography, the Isaacs Kidd Midwater Trawl (IKMT). This net could be towed fast and had no wires obstructing its mouth; it

Left: *Isaacs Kidd trawl being recovered. Peter Foxton is carrying a depth recorder from the net. Dick Burt is wearing the cap. Note the V shaped depressor.*
Right: *Peter Herring washing down a N70V net.*

caught animals of greater size and variety than we had previously caught, but it could not be opened and closed at depth, an essential requirement for our developing interests in the vertical distributions of the ecosystem. Dicky Dobson and Peter Foxton solved this problem by developing the catch dividing bucket, a device at the back end of the net which separated catches from the targeted depth from those made as the net was paid out and recovered. Initially the catch dividing bucket worked by a pre-set pressure switch: later versions operated acoustically.

International Cooperation

In April 1961 Dr Mackintosh resigned as Deputy Director of NIO and Head of Biology to run the Whale Research Unit, which moved back to the Natural History Museum in London in December 1963. Ron Currie replaced him as head of the biology group in June 1962. He brought a new perspective to the research believing that there was a need to link the descriptive ecology of the group more closely to underlying processes, and to expand the expertise available on cruises through the development of strong national and international collaborative links.

This change in leadership coincided with the arrival of the new RRS *Discovery* and the first major programme of the new ship was to participate in the International Indian Ocean Expedition (IIOE), for which George

Deacon was one of the principal proposers. *Discovery* was involved from June 1963 to September 1964. The first leg was a survey of the upwelling region of the southeast coast of Arabia during the southwest monsoon with Ron Currie as principal scientist. The scientific complement comprised nine others from NIO, visiting researchers from other UK laboratories, and five postgraduate students. Two of the latter, Peter Herring and Martin Angel subsequently joined the biology group at NIO.

The first leg was a component of the international programme and also a stand-alone survey of the upwelling along the Oman coast. Five lines of stations running out from the coast were intensively sampled with vertical plankton nets, water bottles and current meters, usually simultaneously. Bathythermographs, bottom grabs, and a few midwater trawls added to the variety as did the neuston net. The latter was designed by Peter David to catch the specialized plankton community (neuston) that inhabits the surface 10 cm of the water column. It consists of a net mounted between the keels of a wooden sled towed from a boom mounted near the ship's bow, effectively sampling undisturbed water beyond the ship's bow wave.[8] Peter Herring recalls an amazing display of bioluminescence one night when the neuston net was launched and caught seven litres of luminous ostracods instead of the usual few millilitres of assorted plankton. They were so numerous that they even sat glowing on the wings of emerging flying fish. The whole experience triggered a lifelong enthusiasm for ostracods in Martin Angel and for bioluminescence in Peter Herring.

The monsoon winds blew almost consistently at near gale force, with the lines of stations aligned across wind and swell. With simultaneous hydrographic sampling from the fore deck, current meters being lowered midships and plankton nets aft it was inevitable that there would be tangles. The nets were particularly vulnerable if the wire strayed aft, because there was a real danger of it getting sucked into the ship's propeller as it did on at least three occasions. Conversely if it strayed forward there was a danger of it tangling with the hydrographic wire. When this happened priority tended to be given to recovering the valuable cast of reversing water bottles at the risk of severing the net throttling rope and losing the net. Most systems broke down at some point–the forward steam winch once failed leaving 3,500 m of wire and water bottles to be wound in by hand, and the seas were often so rough that the accumulators on the aft winch could not cope and, to keep the wire taut, a seaman had to pull constantly on a rope attached to a pulley through which the wire ran!

The first leg demonstrated the presence of upwelling along some 500 miles of Arabian coast with high productivity near the coast reflected in the high numbers of birds foraging there.[9] Sadly all of the plankton samples collected during the IIOE deteriorated because of a failure in the preserving procedure adopted internationally prior to the expedition. As a result a mas-

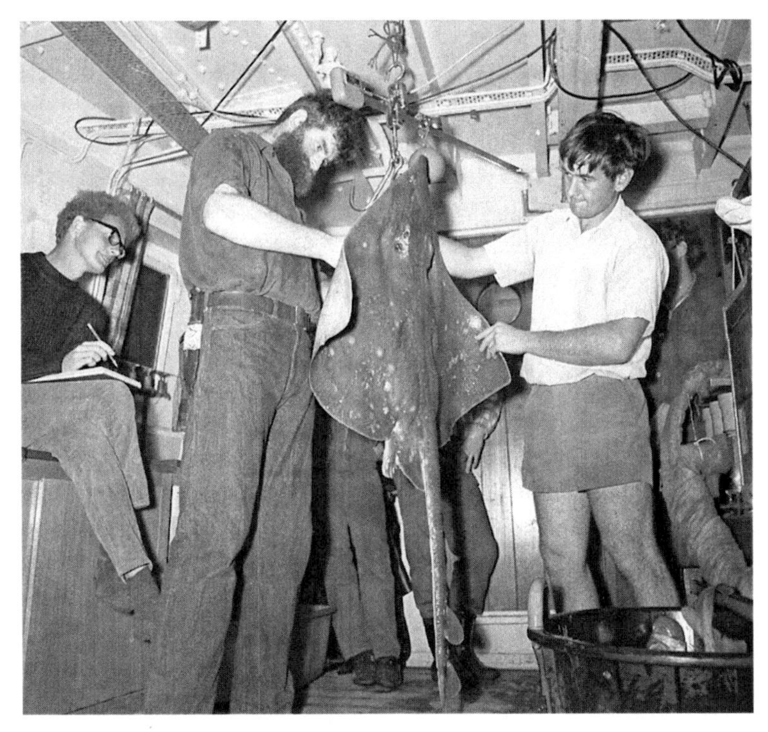

Mike Thurston and Nigel Merrett (white shirt) inspecting a long/bottom line catch. Howard Roe recording the details.

sive body of material collected through this huge international effort was lost and our knowledge of Indian Ocean oceanic ecology severely curtailed.

The second leg of the cruise was mainly concerned with geology and geophysics and all the biologists, apart from the ornithologist Roger Bailey, left the ship in Aden. Martin Angel and Peter Herring went to work at the fisheries laboratory in Zanzibar, where they met Nigel Merrett (later to join NIO) and witnessed independence celebrations and a bloody revolution before rejoining *Discovery* for the third leg. During this the emphasis switched to physical oceanography under the leadership of John Swallow, but routine biological observations continued and samples were obtained throughout much of the western Indian Ocean. After working off the Arabian coast the paucity of plankton in the open tropical waters was a revelation. A simple way of assessing the abundance of phytoplankton in the water is by seeing at what depth a Secchi disc (a standard white disc) lowered into the water disappears from sight. Off the Oman coast this was 2 or 3 metres depth but at one position just north of the equator it reached 51 metres before disappearing.

Station-keeping problems had been experienced off Oman, but these paled into insignificance compared to the difficulties of working in the Somali Current towards the end of the cruise. It was well known that during the southwest monsoon there could be very strong currents in this region, but nothing prepared us for surface currents of 3-4 knots, and on one occasion 7 knots. Intense upwelling generated extraordinary fronts, marked by sudden temperature changes of up to 7°C and long lines of dead pufferfish killed by 13°C water at the surface. Conditions were extreme, and after losing two nets all attempts to sample below 200 m were abandoned. To break the routine, a shore party visited the island of Hasikaya off the Arabian coast to sample seabirds: six Blue-footed Boobies were taken, but the shotgun was left behind and the outboard motor failed, together with a rowlock, so the dinghy was rowed back to the ship with only one oar. Other islands came and went but bad weather prevented any more landings. After a final call in Aden, the ship sailed home via the Red Sea where the recently discovered hot brine pools were successfully sampled. The water bottles came back from a depth of 2,200 m filled with hot (44°C) water so salty that it crystalised when it dripped onto the deck. It was an extraordinary glimpse of the undreamt-of hydrothermal vents that were not recognised until 13 years later, and was a fitting conclusion to a spectacular voyage.

Back in the UK, Martin Angel, inspired by Indian Ocean ostracods, joined NIO to make the study of these planktonic crustaceans the main theme of his oceanographic career. To progress work on the deep scattering layers, Julian Badcock was recruited to work on mesopelagic fishes and Peter Foxton began studying pelagic shrimps and prawns.

The Deep Scattering Layer

The biological group now possessed a suitable suite of sampling techniques together with a broad base of systematic expertise with which to undertake a major study of Deep Scattering Layers (DSLs). DSLs are a ubiquitous feature of oceanic waters: they are layers within the water column from which sound, for instance from echosounders, is reflected. Some of these DSLs remained at a more or less constant depth, but others migrated up at dusk and moved back down again as dawn approached. So almost certainly the DSLs were the result of sound reflected from living organisms, and this sound could be used as a non-invasive method of studying life in midwater. The September October November December (SOND) cruise in 1965 was a major programme conducted on *Discovery* in the sheltered waters off Fuerteventura in the Canary Islands. The aims were to relate the vertical distributions and migrations of animals to the DSLs revealed by sound transmissions of different frequencies in the upper 1,000 m. It was an ambitious multidisciplinary programme linking the ability of the biologists to sample fine-scale distributions of animals with innovative advances in acoustic technology. The support of

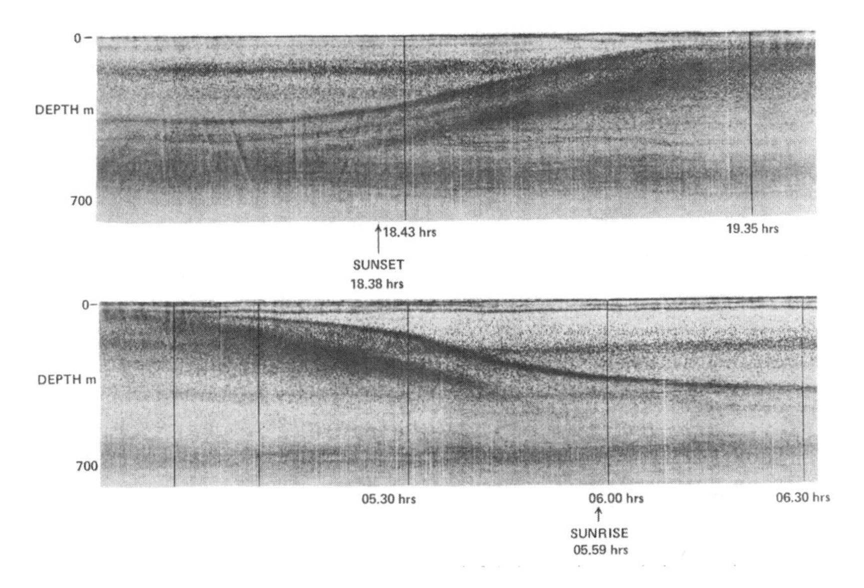

An echosounder record showing three scattering layers moving upwards at sunset and downwards at sunrise. Other layers stay at fixed depths.

our electronics group, notably Brian McCartney and Roy Bowers, and of the engineering group was crucial. Brian and Betty Boden brought underwater light meters from Scripps to record depth profiles of light penetration which were vital for understanding why DSLs occurred at specific depths and what triggered their upward migrations. Expertise in fish taxonomy was strengthened by the participation of Norman (Freddie) Marshall from the Natural History Museum (NHM).

A variety of nets were used to obtain as complete a picture as possible of the full spectrum of pelagic fauna.[10] Micronekton was sampled with the IKMT equipped with the Catch Dividing Bucket; zooplankton with the N70V, and also with a larger conical net fitted with a Catch Dividing Bucket and towed horizontally. The water column was sampled down to a depth of 1,500 m in a series of discrete depth horizons during the hours of both daylight and darkness, providing a comprehensive database from which the bathymetric distributions of the dominant organisms could be determined and related to measurements of underwater irradiance and the movement of the DSLs. In parallel with the biological sampling, acoustic studies determined the target strengths of larger individual animals and the volume scattering at different frequencies. A range of environmental data was collected including light, which was measured at selected depths and continuously at the surface.

The cruise generated a unique and substantial collection of multi-disciplinary data. The laborious task of sorting the biological collections was completed by August 1966, and identification of the specimens by

***Above** and **facing page:** A Catch Dividing Bucket attached to a large conical net. Inside the Y shaped aluminium tube is an acoustically operated flap. The flap is switched when the net reaches the desired depth and re-switched before hauling the net in. This gives a catch from a known depth in one 'trouser leg' and a catch from the deployment and recovery in the other.*

the in-house specialists began. Ten groups, which made up the bulk of the organisms sampled, were eventually studied; fish,[11] cephalopods, salps, euphausiids, decapod crustaceans, ostracods,[12] copepods,[13] chaetognaths, amphipods and siphonophores. The vertical distributions and migrations of these groups were described in unprecedented detail and direct relationships between biological distributions, community structures, *in situ* light and sound scattering layers were established for the first time. Taken together the SOND papers comprise one of the first-ever holistic studies of oceanic communities, and the results stimulated many of the future ocean biology programmes at NIO and abroad.

Ron Currie on Discovery *during the IIOE.*

1966 brought further significant staff changes. Ron Currie resigned as Head of Biology to become Director and Secretary of the Scottish Marine Biological Association laboratories at Millport and Oban. He was succeeded in the New Year by Peter David. Since Ron Currie intended to continue his interests in primary

productivity and phytoplankton biochemistry at SMBA, he and Peter David mutually agreed that NIO biology should continue to focus on the ecology of oceanic zooplankton in the North Atlantic, a task for which the group was now particularly well fitted following its expansion under Ron's leadership. Peter Herring, one of the Indian Ocean students, joined to continue his studies on the pigments of pelagic animals and began his work on bioluminescence.

New nets, new initiatives

Peter David brought a formidable intellect, sound judgment and a firm commitment to continue and develop the general direction of the research. He devoted his time fully to managing the group and its science rather than pursuing his own research, which enabled his team to flourish. He remained head of the group until his retirement in 1982 when he was succeeded by Martin Angel.

The analysis of the SOND data had revealed that the Catch Dividing Bucket caused problems with some of the catch being held up within the main body of the nets. The problem did not invalidate the conclusions of the SOND cruise, but it had to be solved before we could determine how many and what biomass of each species occurred at each depth. The challenge was taken up by Malcolm Clarke and Arthur Baker with help from the engineering and electronics groups.[14, 15] They designed a completely new system consisting of a pair of nets the mouths of which opened and closed at depth. The system became known as the Rectangular Midwater Trawl (RMT1+8) with a small 1 m^2 net for plankton fished together with an 8 m^2 net for micronekton. The nets were opened and closed via a release mechanism, operated by acoustic signals transmitted from the ship. A net

Left: *Malcolm Clarke with an early RMT 1+8 net monitor and release gear.*
Right: *Deploying the RMT 1+8 with open nets.*

monitor was mounted above the RMT1+8 which continuously acoustically transmitted the depth of the net, state of opening/closing and (later) environmental parameters such as flow, light and water temperature. All these data were recorded in real time on a modified Mufax recorder aboard the ship. The first trials of this new system in 1967 were so successful that the IKMT was abandoned, and the RMT and its subsequent modifications became an international success, enabling controlled sampling down to depths in excess of 5,000 m, and with this a huge expansion in our ability to catch and describe many novel species and to quantitatively study deep ocean communities from the surface to the seabed.

Malcolm Clarke's quest to improve sampling of oceanic squid continued in a succession of trials with a variety of large midwater trawls including the commercial Engels Trawl, ingenious but impractical 'pop-up' nets, and larger and larger versions of the RMT. These did increase the size of the animals caught but the samples were disappointingly small, tending to confirm the advice of Sir Alister Hardy that to sample oceanic squid properly one needed a net the size of the Royal Albert Hall, prophetically anticipating the future use of vast purse seines and drift nets in commercial deep-sea fisheries.

The SOND cruise created a collective momentum and interest in studying the vertical structure of pelagic ecosystems which Peter David was keen to develop into a major new programme. The earlier work of Stanley Kemp (the first director of the Discovery Investigations) on the decapod genus

Acanthephyra indicated that there was a geographic succession of shallow and deep species from the South to the North Atlantic, possibly associated with the main features of the ocean circulation. Our new programme aimed to investigate any such changes by examining the vertical structure of deep-sea communities on a transect along 20°W at approximately 10° intervals from the equator to 60°N. These stations were sampled in successive years from October 1969 to May 1972 during which time the taxonomic breadth of the biology group expanded with the arrival of Howard Roe from the Whale Research Unit to work on copepods (thereby downsizing from the largest to some of the smallest marine animals); Mike Thurston from the British Antarctic Survey for amphipods; Nigel Merrett for fish and Phil Pugh for siphonophores. Results from this transect have shown geographical changes in biological distributions and community structure linked to the physical oceanography, and work on the samples continues to this day.

The broader geographical coverage of the sampling gave Peter Herring access to a wider selection of species with which to advance his now-classic studies of bioluminescence and pigmentation of oceanic animals.[16] Peter's work was greatly enhanced by the constant temperature laboratory with its clean sea water supply onboard *Discovery*, making it possible to keep specimens alive in controlled conditions, albeit for a limited time, and to undertake behavioural and physiological studies. This facility, together with our taxonomic and sampling expertise and knowledge offered a unique opportunity for others to join our cruises and work with freshly caught deep-sea animals. Strong collaborative links developed with numerous colleagues from laboratories and universities within the UK and abroad, and these cooperative links continue to expand to this day.

Mike Fasham joined NIO in September 1968. He came from a geophysics background to work with Jim Crease to build up the computer facilities, but was soon inveigled by Martin Angel into analysing the SOND data using some of the new non-parametric techniques that had just been developed, including cluster analysis, principal components analysis and later factor analysis.[17] Mike's analytical skills became fundamental to our attempts to understand the mechanisms underlying the variability in time and space of the communities we were studying. Post-NIO he became one of the most influential biophysical modellers in the world and his involvement in large-scale international research programmes enhanced the science of these and facilitated our interactions with them.

Interest in vertical migrations continued. The ability to deploy, fish, recover and re-launch the RMT1+8 rapidly meant that series of samples could be collected quickly enough to follow migrations. In 1972 a series of samples was collected over 24 h at a depth of 250 m, which was worked up as a collaborative effort analysing the movements of over 100 species belonging to nine groups of animals.[18] This work demonstrated the continuously changing nature of deep-sea pelagic communities with time, and also allowed the first detailed examination of the feeding patterns of a variety of deep-sea animals, and it expanded into much larger programmes post-NIO.

Throughout the NIO years there had been little serious attempt to study bottom-living animals. This was recognised as a major hole in our research and Mike Longbottom was appointed in 1969 to develop a benthic programme linked to the ongoing pelagic work. Sadly Mike died a few years later and the benthic biology programme that subsequently thrived did not get going until after NIO transformed into the Institute of Oceanographic Sciences in 1973.

Postscript

The Discovery Investigations were set in the context of relevance to commercial whaling, focussing on the need to gain a proper understanding of the biological structure of the Southern Ocean ecosystem and how it is driven by physical and chemical processes. This general approach was carried forward by the work of the biology group throughout the NIO years, during which the emphasis moved from classical studies of individual species' distributions to more quantitative community studies and multidisciplinary research. These changes would not have been possible without the development of quantitative sampling gears and new technologies, and emerging understanding of physical and chemical oceanography. Both areas were directly enhanced by our being part of a multidisciplinary institute. The move to Wormley was a seminal event for the biologists, the significance of which cannot be overstated. We were privileged to work within a multidisciplinary institute at a time when curiosity-led research and in-house technical innovation was not only funded but positively encouraged by a director whose philosophy was to let his staff get on with their research while he dealt with the often unpalatable aspects of administration. Such an enlightened approach became untenable as central government changed the funding of science, took greater control of research, and introduced the customer/contractor philosophy to science funding. Funding increased but support for long-term research became unfashionable and the ethos of the institute post-NIO changed. The glory days were over.

The decision to focus on the deep oceans beyond the continental shelf was based on the firm belief that this ecosystem is of global importance. The resultant biological research has proved relevant to many of the applied problems that later emerged. Our results gave us some of the answers to environmental concerns associated with deep-sea disposal of pollutants, deep-sea mining, oceanic fisheries, and, more recently by climate change and the need to conserve biodiversity. If we did not already know the answers we had the expertise and know-how to provide them, although it was often difficult to get 'contractors' to meet the full cost of the research. A lesson that NIO clearly demonstrated was the added value provided by a professional pool of multidisciplinary expertise that could be switched to deal with emerging issues. Examples include the assessment of the possible impact of the disposal of high-level radioactive waste in the deep-sea and the disposal of redundant oil production platforms. One of the interesting results from the 1970s transect data is that the (now) well-known reduction in species richness that occurs from the tropics to polar latitudes is not restricted to the upper surface layers but occurs at all depths throughout the water column. The NIO biological legacy is extremely positive, and available in the published papers, reports, databases and web-sites generated by a team of expert and dedicated researchers. Today these data have the potential to be a baseline against which changes in oceanic communities resulting from, for example, climate change could be assessed. Much of the systematic expertise that was nurtured at NIO has been lost as new areas of research and different specialities have been accorded priority. The trend away from descriptive ecology and taxonomy, which is not restricted to the UK, is unfortunate at a time when tracking and understanding changes in oceanic communities and processes are becoming critical.

6

Whales and whaling

Howard Roe

Discovery Investigations prelude

Research to understand the biology and population dynamics of the great whales (mainly Blue, Fin, Humpback, Sei and Sperm) and their role in the Antarctic ecosystem was central to the creation of the Discovery Committee and subsequent Discovery Investigations in the 1920s. Antarctic whale research by the UK was essentially set up to provide scientific information to help in the regulation of the whaling industry funded by a tax on whale oil processed in South Georgia. This rationale for the work and for continuing with it was reiterated by the Natural Environment Research Council (NERC) in the Council's second annual report for 1967-68, although by then the UK no longer had a whaling industry. Arguably this basis for whale research is also a very early example of what became known as the customer/contractor principle put forward by Lord Rothschild in 1971 and which changed the face of scientific funding in the UK: it was research funded by industry for the benefit of the industry, the research community and ultimately the populations of whales. It is ironic that this tax on whale oil contributed to the creation of pelagic whaling from factory ships with no need to call into the Falklands, and the resultant over-depletion of whale stocks. The first of these ships, the Norwegian *Lancing*, worked in the 1925-26 Antarctic whaling season; from that point, factory ships and their fleets of catchers dominated the industry and the subsequent destruction of whale stocks for much of the rest of the century.

By the time that NIO was created in 1949 whaling research was a small component of the total research effort. The only member of staff specifically employed for this was Dr Robert Clarke – appointed in 1949 – with Dr Helene Bargmann and Dr Neil Mackintosh providing help and advice. Dr Mackintosh had been Head of the Discovery Investigations and was then Deputy Director of NIO and Head of Biology. He was a renowned expert on whales and whaling and more generally on the Antarctic, and remained a pivotal figure in whale research until his death in 1974. Throughout the lifetime of NIO he published a range of papers on whale distributions and migrations, on stocks of whales and the whaling industry, and on plankton and Southern Ocean ecology.[1,2] Dr Bargmann was curator of the Discovery Collections which were housed in the Discovery Hut at the Natural History Museum in London, also home to both her and Robert Clarke together with other non-whaling biologists. Part of her job was to examine the ovaries from Blue whales collected at whaling stations and factory ships as part of a research programme to work out the breeding cycle of Blue whales, which were the mainstay of the pre-war industry.

Whale research themes

Understanding the life histories and reproductive cycles of the great whales remained a central theme for most of the next 50 years and the results underpin our present knowledge of whale population dynamics. Research was initially based mainly upon examining ovaries, which retain a record of an individual's reproductive history because scars of previous ovulations and pregnancies remain within them. These scars are known as *corpora albicantia* and counting them gives a way to estimate recruitment rates and age, both essential parameters for stock assessments. Other methods of age determination were found later, and in the years to come NIO scientists worked out the reproductive cycles of Fin, Sei and Sperm whales together with accurate ways of determining their age.

The other major research theme, which continued throughout the time of NIO and beyond, focussed on the seasonal migrations and associated behavioural changes exhibited by the great whales. Essentially baleen whales and Sperm whales migrate between polar regions and the tropics/subtropics, feeding in cold water and giving birth in warm water, where they do not feed. There are specific differences in the range and timing of migrations. Within species there are different stocks of whales that follow different migration routes, particularly in Humpback whales, and there are sexual differences in movements and behaviour especially in Sperm whales. This knowledge, and much more, was determined by many researchers over many decades and depended upon observations of the times at which whales were caught in particular areas, on observations from ships, and from the results of whale marking.

Whale marking was established before the Second World War. Marks are metal tubes about 25 cm long with a solid conical head, each has a unique number engraved upon it. They are fired into surfacing whales from a modified shotgun, embed themselves beneath the blubber, and are recovered when the whale is subsequently caught and processed. The date and position on marking and subsequent capture provide direct

Whaleboats at Fayal, Azores, 1949 (Discovery Report XXVI).

evidence of the movements of the whale. NIO biologists made significant contributions to understanding the migrations of the great whales and their dispersals on their feeding and breeding grounds. Many years later marks also played a fundamental role in the search for a method of absolute age determination, because over the years many marks were recovered from whales that had been marked decades previously, and the certainty provided by the mark and re-capture dates could be used to verify or refute other methods of age determination.

Whale research and NIO

Robert Clarke sailed on the RRS *William Scoresby* in 1950 to survey the Benguela Current and to mark whales on their winter grounds in the subtropics. He led the subsequent voyage in the Indian Ocean but the proposed crossing to Australia was abandoned because of bad weather and only a few Humpback and Sperm whales were marked off Madagascar before the *William Scoresby* returned to London on what proved to be her last voyage. During the previous year he had spent the summer working on the Azores where open-boat Sperm whaling using hand thrown harpoons in the style of Moby Dick was still practised. Robert made a ciné film of both the hunt and subsequent processing which is a graphic testimony to an heroic way of life immortalised by Melville but which no longer exists.

Harpooning a 53 ft bull Sperm whale off Fayal 1949 from a ciné film (Discovery Report XXVI).

His subsequent report on the open boat whaling is a classic account,[1] and his knowledge of the techniques and way of life of the old Yankee whalers lead to his becoming an adviser for John Huston's film *Moby Dick*. In addition to filming, his records and collection of samples from Sperm whales on this and future trips enabled him to make significant contributions to the biology of the species, including observations on ambergris, once worth more than its weight in gold as a 'fixative' for perfumes, and a rare deep-sea angler fish and a giant squid found in (different) Sperm whale stomachs.[1,3,4] This last was a forerunner to future NIO research making use of Sperm whales to sample oceanic squid.

The primary body responsible for regulating the whaling industry is the International Whaling Commission (IWC) established in 1946. From the outset biologists from NIO played a leading role in the Commission and its committees. Dr Mackintosh was the Chairman of the IWC's Scientific Committee from its first meeting until 1963. This committee is responsible for making reports to the Commission on the state of whale stocks and recommendations on conservation measures. Over the years all of the NIO whale research biologists participated in the Commission and provided advice to this and to the UK government.

Whale research depended mainly upon the collection of material from carcasses as these were processed at whaling stations or onboard factory ships. Collections were made mostly by NIO biologists or by whaling

Heaving up a 53 ft bull Sperm whale at the whaling station at Fayal, 1949 (Discovery Report Vol XXVI).

inspectors appointed for the season who collected samples and made observations as part of their duties. The samples were subsequently sent to England for analysis. All of the NIO whale biologists spent time on whaling platforms, some on many occasions and at many different places, collecting from a number of Antarctic factory ships, from stations on South Georgia, from Durban and Saldanha Bay in South Africa, from the Azores, Madeira, the Hebrides and Iceland in the North Atlantic, and from Chile and Peru in the SE Pacific. To these direct collections were added material from whaling inspectors, and the collections made by whaling companies. In these ways a very large amount of material was collected and analysed, providing much of our understanding of the biology and population dynamics of Blue, Fin, Humpback, Sei and Sperm whales which ultimately led to the regulation and restrictions on whaling that exist today.

Sidney Brown joined NIO in 1950 and spent his whole career involved in whale research, becoming one of the most respected and knowledgeable whaling biologists in the world. He started working on the distributions of Blue and Fin whales from marking and observations at sea, and shortly afterwards established the international whale marking programme in the Antarctic which he co-ordinated and reported on annually for the next 35 years or so.[1,5] Sidney also started a voluntary observing system, co-ordinated by the UK Meteorological Office, which provided much useful data on the occurrence of whales in the world's oceans for many years until it was replaced in 1965 by another scheme co-ordinated by the Scientific Committee on Antarctic Research (SCAR). Sidney worked on an Antarctic factory ship, the *Baleana*, and on whaling stations in the Antarctic and Iceland. He took part in observational cruises, mentored, educated and encouraged young biologists, and initiated programmes on Southern Right whales, on cetacean strandings in the UK, on N. Atlantic Pilot whales and on observations from yacht races. His publications are diverse, including such gems as records of swordfish attacks on whales.[6]

Robert Clarke and Sidney Brown were joined in 1953 by Richard (Dick) Laws, a biologist from the Falkland Islands Dependencies Survey who had worked on the biology and reproduction of Antarctic seals. He stayed until moving to the Nuffield Foundation of Animal Ecology in East Africa in 1960, and had an extremely distinguished career researching the reproduction and biology of various large mammals including Elephant seals, Antarctic Fin whales, hippopotamus and elephants, becoming the Director of the British Antarctic Survey and Master of St Edmund's College, Cambridge. He spent the 1953-54 season aboard the *Balaena* and devoted most of his time at NIO to elucidating the reproductive cycle, growth and age of Antarctic Fin whales from ovaries and the accumulation rate of *corpora albicantia.*[1] His papers remain the most comprehensive studies on the growth and reproductive cycles of these whales.

NIO was established at Wormley in 1953 and the whale biologists moved there from London. Work continued on various fronts, but one of the most influential discoveries for whale research and subsequent stock assessments was made by Francis Fraser and Peter Purves of the Natural History Museum during their work on hearing in baleen whales. They looked at a curious structure in the external auditory meatus of the whale, an apparently waxy plug linking the inner ear to the outside, which in a large whale can be more than a metre on each side! This ear plug had been noted in 1910 but ignored until Fraser and Purves found that it was an extremely good conductor of sound. This in turn prompted them to examine it in detail and they found that it was not an amorphous lump of wax but composed of keratin with a distinct structure of alternating light and dark layers at its base. Purves subsequently concluded that the layers could be related to age, and that two pairs of layers were likely formed each year.[1] This was a critical discovery. It seemed that ear plugs might provide a means of getting an absolute age for individual whales and hence the age structures and trends of populations vital for accurate stock assessments. Following this discovery, collection of ear plugs became an essential part of the sampling programme at the various platforms, and over the next few years thousands of plugs from Fin and Sei whales were collected all over the world, many of them finishing up at NIO where routine counts of numbers of laminations were correlated with numbers of *corpora* in ovaries and numbers of ridges on baleen plates; another way of ageing young whales. Sperm whales do not have ear plugs, but in the 1950s it was found that their teeth have layers in the dentine, and so these were routinely collected. But it was many years before the formation rate of these layers was established.

Malcolm Clarke (not related to Robert Clarke) spent the 1955-56 season collecting material on board the factory ship *Southern Harvester* where he met Neil MacLeod who was working on the ship. Malcolm joined NIO in 1958, worked initially on squid remains from sperm whale stomachs and went on to have a very distinguished career as an expert on oceanic cephalopods.[7] His interest in Sperm whales continued throughout his career; in 1970 he described the role of the spermaceti organ (the large oil-filled sac in the head of these whales) in diving and buoyancy,[8] and in his retirement he has opened a whaling museum on the Azores. Neil MacLeod joined NIO in 1961 and initially prepared tooth and ovary samples for analysis and collected material in Durban. He then began working with Malcolm Clarke and became an expert on identifying squid beaks from Sperm whale stomachs, on which he worked for the rest of his career. Other work on squid remains started at this time when Dr Anna Bidder from Cambridge University began to study the giant cranchids from Sperm whale stomachs that were stored in the Discovery Hut.

Whaling factory ship Southern Venturer *in the Antarctic. Note the stern opening and slipway up which the whale carcasses were hauled to be processed on deck.*

A major change in personnel occurred in 1958 when Robert Clarke was seconded to the UN Food and Agriculture Organisation to begin a programme of whale research in Chile, Peru and Ecuador. This secondment took him to South America where he continued to work for the rest of his career. With his co-workers he produced the definitive works on the biology and stocks of whales (principally Sperm whales) in the SE Pacific.[9,10]

The 1960s saw several further changes in people and significant new activity and research. In the Antarctic season 1960-61 David Crisp and John Bannister worked as inspectors aboard the *Southern Venturer* and on South Georgia respectively. Both subsequently joined the staff. Crisp stayed for one year before joining the Nature Conservancy, but wrote a paper on the biomass and weights of whales relative to pre-war catches in the course of which he questioned the use of the Blue Whale Unit; the standard method of catch regulation whereby one Blue whale equated to two Fin whales and six Sei whales.[11] The Blue Whale Unit was a very crude measure that did not distinguish between different species and stocks and therefore did not allow protection of these, but it was not abandoned as a regulatory tool until many years later.

In 1961 the International Whaling Commission recognised that an international approach was needed to maximise understanding of whale stocks. It was agreed that countries should pool data on recruitment, mortality and catch statistics and an ad hoc group of specialists in population dynamics from countries not involved in Antarctic whaling was set up to analyse the resultant data set. NIO was the largest single contributor of data to the group, providing information on some 11,000 whales. This initiative resulted in the recognition that the stocks of Blue and Humpback whales in the Antarctic were so low that they needed total protection for an indefinite period. Fin whale stocks were also low enough to need a reduction in catch of about 25%. These recommendations resulted in total protection for Antarctic Blue and Humpback whales in 1963. Fin whale catches were reduced, but remained too high for stock recovery; catches of Sei and Sperm whales, however, increased.

John Bannister left in 1964 to become Director of the Western Australian Museum in Perth, Australia, but during his time at NIO a major research programme based on whales taken at the South African stations at Durban and Saldanha Bay began.[12] This work generated important information on Sei and Sperm whales, the catches of which were increasing in response to falling catches of the larger whales, and also provided critical earplug material from Fin whales during the southern winter. These earplugs were to provide the key to unlocking the riddle of age determination in baleen whales.

The South African work was continued and significantly developed by Ray Gambell who joined in 1963 and spent many years working on all

Fin whale being taken by rail truck from the slipway to the whaling station at Durban.

aspects of the (then) poorly known Sei and Sperm whales taken off Durban. He eventually wrote the definitive accounts of the biology and population dynamics of both species and played a major role in establishing methods for determining their ages.[13,14,15] He was extremely active in developing catch limits, in whale marking and, apart from his main interest in Sei and Sperm whales, wrote papers as diverse as a description of a pygmy Blue whale from South Africa and commensal copepods from the baleen plates of whales. After Dr Mackintosh retired in 1968 Ray became head of the Whale Research Unit until 1976, when he left to become Secretary of the International Whaling Commission.

The last British factory ship, the *Southern Harvester*, worked in the 1962-63 Antarctic season with Nigel Merrett as the whaling inspector. In addition to the normal collections of ovaries, earplugs, stomach contents and so on, Nigel made observations on the abundance and tastiness of ice fish. These now support a commercially viable fishery, and Nigel subsequently joined NIO in 1968 and became an international expert on deep-sea fish. Sidney Brown was on South Georgia for the same season where the two remaining land stations were by now leased to Japanese companies.

In 1963 the whaling biologists moved back to London, and the Whale Research Unit (WRU) was created and housed within the old Discovery Hut. Dr Mackintosh was head of the WRU, with Sidney Brown, Ray Gambell and (briefly) John Bannister. Robert Clarke was in South America. The next recruit to the WRU was Howard Roe who arrived in 1965 with the task of trying to establish once and for all the validity

Humpback whale being flensed at the whaling station at Durban.

of using earplugs as a means of absolute age determination. By now it was apparent from old whale mark returns that the rate of formation of laminae could not be two a year; at least throughout the life of the whale. The answer came from studying the formation of the different layers in plugs taken from whales caught at different ends of their migration and feeding routes, from the Antarctic and South Africa. By doing this Roe showed that in Fin whales one pair of laminae was formed each year, a dark layer in the winter and a light fatty layer in the summer when the whales were feeding. For the first time a means of determining the absolute age of the Fin whale had been found and this discovery formed the basis of all subsequent stock assessments.[16] Gambell, together with Christine Lockyer, later established the same annual rate for Sei whale plugs, and also for the layers in Sperm whale teeth. Hence by the end of the 1960s/early 1970s each of the large whale species that were still caught commercially could be reliably aged. Christine Lockyer (as Christine Grzegorzewska) worked at the WRU as a vacation student in the summer of 1967 before her marriage. She joined the staff in 1968 and later (1972) showed that the age of sexual maturity in Antarctic Fin whales had declined in response to decreasing populations from 11-12 years old in the early 1950s to between 6-7 years in the late 1960s.[17]

The IWC reduced the catch quota for Fin and Sei whales to below the combined sustainable yield for the two species for the 1967-68 Antarctic season. The quota was then further reduced in subsequent years and in 1972 the Blue Whale Unit was abandoned in favour of setting quotas for individual species and stocks. Antarctic Fin whales did not receive total protection until 1976 and Sei whales and Sperm whales not until 1979, when commercial whaling in the Antarctic ended except for the much smaller Minke whales.

Whaling from South Georgia ceased in December 1965 and a new source of material was established at the whaling station in Iceland in the summer of 1967 by Sidney Brown and Howard Roe. This provided comparative material on North Atlantic Fin, Sei and Sperm whales, and it was discovered that the Sperm whales caught there uniquely feed mostly upon bottom living fish instead of the squid that form their diet elsewhere. These whales were caught in water depths of between 500-2,000 m and their stomach contents provided further confirmation of the great depths to which Sperm whales dive.[18] Howard Roe moved to the main NIO biology group at Wormley in 1968 and spent his career working on plankton dynamics, sampling systems and biophysical interactions before co-ordinating the development of the Southampton Oceanography Centre in the late 1980s and subsequently becoming its Director. Sidney Brown made two further trips to Iceland in 1969 and 1973.

The most momentous change in the history of NIO since its establishment in 1949 occurred in 1966, when the Natural Environment Research Council was established and took over the ownership of NIO from the National Oceanographic Council. In its second annual report of 1967-68 NERC noted that, although the UK no longer had a whaling industry, the WRU should "*continue its research ... because of the importance of the whale in the marine ecological system and to prepare for the time when the stocks of whales have again increased to the level of being commercially exploitable*". And so it did. Work continued as described on the stocks of whales taken off South Africa and Iceland, together with comparative studies of population changes with time in Antarctic stocks. A NERC Visiting Group inspected the WRU in 1972 and concluded that the group would benefit by closer association with other scientists working in similar fields. In 1977 this came to pass when the WRU was disbanded as an entity and joined forces with the NERC group working on seals to form the Sea Mammal Research Unit, based initially in Cambridge and currently at St Andrews University. By now NIO had also ceased to exist: it became the Institute of Oceanographic Sciences in 1973, and the direct link that had existed since the 1920s between ocean ecology and the great whales was broken.

What was it really like?

The preceding account summarises the evolution of whale research at NIO and the comings and goings of the associated staff who made such massive contributions to our understanding of the great whales and ultimately to their conservation. Their contribution was all the more remarkable because there were never more than a few people involved at any one time, and it is also worth noting that several very distinguished individuals began their careers with involvement in whales and whaling research before later making their mark in other areas of marine science. The research and activities outlined here cannot be repeated. There is no longer the opportunity or the industrial basis to support it. So what was it like to take part in whale research? What was it like to work on a whaling platform? To try to give some idea, the remainder of this account draws upon my memories of working at the WRU between September 1965 and December 1967.

At that time, the WRU was housed in the Discovery Hut at the Natural History Museum in London. My appointment was bizarre by today's bureaucratic standards and my job interview was conducted on the telephone via a public call box with the long-distance call charges reversed to Dr Mackintosh. To get to the hut involved walking through the mysterious basements of the Museum, with stuffed animals, skeletons, lumps of rock and cabinets at every turn, before emerging into the daylight at the rear of the Museum. The hut was by then showing its age, a large black, wooden building set on brick piles with tanks of formalin containing squid remains arranged about the door. Inside were three laboratory areas, an area where the Museum's coral collection was housed, and a large central space off which opened several offices. Along one side of the central hall were rows of large cupboards housing the Discovery Collections, thousands of jars with preserved krill, whale bits, and squid beaks, together with cupboards full of log books, punched data cards, old whale marks, cruise reports and records of meetings. Everywhere was pervaded by a curious smell, part whale oil, part formalin and (no doubt) part age. Days when teeth were being prepared added the distinctive smell of burning bone.

There were separate offices for myself, Ray Gambell, Sidney Brown, Susan Hiddleston – who acted as secretary and general factotum for all of us – and Dr Mackintosh. Dr Mackintosh's office was of course the biggest, and was filled with a large table covered with books, charts and papers and bearing a big yellowing globe, a set of chairs for the table that were rarely used, his desk and to the side of this a low armchair. All this furniture was made of some dark tropical hard wood and was polished regularly by Susan. Behind the desk was a large glass-fronted cabinet full of specimens but also containing exquisite models of whales and whale boats carved from bone and ivory.

Dr Mackintosh was extremely formal, but also extremely kind to this young recruit. He was always referred to by everybody as Dr Mackintosh, and he always called men by their surnames only, even colleagues with whom he had worked for decades. Jackets and ties were obligatory, and if

The whaling station in Iceland; a Fin whale on the slipway.

you went to see him in his office you would sit in the armchair at the side of his desk whilst he looked down on you from his central vantage point. Twice a day Susan would call us to tea and coffee, when we gathered around a large table at the end of the central space. Dr Mackintosh would puff away at his cigarettes, invariably directing the smoke upwards to the ceiling out of the corners of his mouth, and we talked about work, future plans, the Discovery Investigations and the heroic age of Antarctic exploration and research. Hanging on the yellowing walls were old black and white photographs of *Discovery* and the whaling platforms at South Georgia, and on the end wall overlooking the table was a Narwhal tusk donated by the Canadian High Commissioner after the war; this is now in the National Oceanography Centre's library in Southampton. Visitors were frequent guests at tea time: I recall old Discovery Investigation colleagues including (Sir) Alister Hardy and Francis Fraser, (Sir) Peter Scott with his Antarctic interests, Peter Purves who worked in a Museum laboratory opposite the hut and, in the summers, Anna Bidder who came down to measure the giant cranchid squid housed in the tanks by the door. Anna would preside over these proceedings wrapped in a white plastic sheet whilst one of our assistants hauled the squid onto a table where they were measured to check on their shrinkage in formalin before being returned to their smelly tanks.

Ray Gambell and Sidney Brown were very friendly and very willing to advise, teach and help a newcomer. Olga Nash, Michael Harry and later Ian Keevil assisted us all and lived in the laboratories together with the tools of the trade (no Health and Safety concerns here!). These included a large bacon

A whale catcher bringing a Fin and a Sei whale to the station.

slicer such as used to be found in every corner shop but here used for slicing ovaries, a bandsaw for cutting teeth, large sinks, vats of formalin, and jars, barrels and sacks of specimens everywhere. Olga, Michael and Ian sliced ovaries and made *corpora* counts, cut teeth in half and then etched the cut surfaces to enable paper 'brass' rubbings to be made of the growth layers, prepared earplugs for counting laminations, made histological preparations and generally kept us all going. The hut was a good place to work–even the cockroaches seemed content. I stayed two happy years there, spending most of my time cutting frozen sections of earplugs accompanied by clouds of frozen carbon dioxide before examining the results and using an old handle-driven paper roll calculator to do any sums.

Samples were sent to the hut from around the world, and because there was no longer a UK whaling industry I went to Iceland in the summer of 1967 to collect material from the Fin, Sei and Sperm whales caught there. Sidney Brown accompanied me for the first two weeks and taught me the fine art of large-scale butchery. Working there was unforgettable; the sights, sounds and smells were a continuous assault on the senses. Everybody was very friendly and helpful (virtually all spoke excellent English), and very curious to know what I was doing! My novelty value was such that I was introduced to the (then) Crown Prince Harald of Norway who visited the whaling station and just missed being showered by an exploding whale that had been kept unprocessed to ensure that there was something for him to see.

The whaling station was (and still is) at the end of a picturesque fjord, built into the side of a cliff of purple basalt which overlooks the flensing

Flensing a Fin whale in Iceland.

platform and which housed colonies of seagulls. These descended onto the platform at every opportunity to gorge on the remains of whales, and dodging the rain of droppings when returning to the platform after a break was an occupational hazard! The platform was built on top of the processing factory, as in a floating factory ship, with cookers opening flush into the platform which was surrounded on the sides with winches, bone saws and a meat hut. At the seaward end a slipway led from the sea to the platform, and at the opposite end a hut contained a large winch for hauling up the carcasses. Here were sited the rengi tanks. These last were vats of sour milk pickling the 'tender' strips of grooved throat blubber of Fin and Sei whales into a local delicacy.

When working, the whole area was festooned with clouds of steam from the winches and saws. Piles of blubber, bone, guts and meat were heaped up waiting their turn in the cookers; dozens of men hauled bits of whale, cut meat and bone, and pushed sheets of blubber into the cookers. Cables criss-crossed the wooden platform which was covered with blood, grease and sea water from hoses which continuously washed the platform in the vicinity of the meat hut and lubricated the slipway to ease the hauling up of the whales. The plan foreman and the flensers constantly called out instructions to the winch operators; the elderly machinery clanked, banged and hissed; ripping, tearing noises filled the air, punctuated by dull thuds as the whales were turned over and tons of muscle and bone rolled and slithered across the platform. Joints of meat were cut from the freshest baleen whales and sent off to the kitchens - we ate it every day and it was delicious! The scene was bloody, but absolutely compelling, and the operation extremely efficient. Whales were completely processed within 1-2 hours, slow by

Sawing up the head of a Sperm whale in Iceland. Note the white spermaceti pouring out of the head; this high-quality wax was formerly used for making church candles.

Antarctic standards because of the meat processing but quite fast enough for a novice like myself. Work continued round the clock when the whole scene was lit up and clouds of steam rose into the night sky.

Whales were caught in the Denmark Strait between Iceland and Greenland and towed to the station by the catcher boats, which shackled the carcasses alongside, two Fin or Sperm whales and up to six Sei whales at a time. The whales were delivered to a long jetty at the foot of the slipway, pulled to the bottom of the slip where the main winch cable was attached to the tail of each carcase in turn. Each was then hauled up the slipway onto the platform. As the carcass came over the lip of the slipway the flensers started to cut along the body with their razor-sharp flensing knives, shaped like hockey sticks. Flaps were cut in the blubber, cables attached and the blubber peeled back like a banana skin accompanied by flensers cutting it free with loud ripping noises. The lower jaw was then pulled off. In baleen whales this stood out like a huge rounded V before being hauled to the bone saws, exposing the balloon-like grey and pink tongue, and the upper jaw with its rows of baleen plates. In Sperm whales the lower jaw is rod-like and lined with two rows of conical teeth, there is no large tongue and the upper jaw and throat area are pure white in colour.

Next the main body of the whale, the meat, was cut into individual blocks by the flensers climbing onto the carcass and the blocks swung into the meat hut by an overhead hoist; the carcass was then rolled over, and the remaining blubber and head cut off before the rest of the meat was removed. The rib cage was cut off, when the guts and internal organs spilled out, a colourful explosion of purples, whites, greys and reds, loops

Howard Roe weighing a 23 cm Fin whale foetus.

and coils and huge bags of stomach, heart and lungs. Everything except the baleen disappeared into the cookers or meat hut where it was rendered into oil, meal and frozen meat which, when I was there, was exported to the UK where it fed our pet dogs, an ignoble end for such magnificent animals.

During the flensing process I was busy! Each whale was identified, numbered, sexed and measured. Any unusual colour patterns were noted, and for baleen whales the presence or absence on the skin of a greeny-yellow diatom film which gives a measure of the time spent in cold water. On the heads of Sperm whales the occurrence of scars caused by the suckers of squid were recorded and measured. The thickness of the blubber was measured, and in females the thickness of the mammary glands. Ear plugs were cut out from the base of the skull of baleen whales as this was being sawn up: teeth were cut from both the lower and upper jaw of Sperm whales. When the internal organs were exposed either complete pairs of ovaries were collected, pinky-grey and knobbly, or testes, like pink bolsters, were sliced up and sampled. Stomach contents were noted, sampled and sometimes weighed, and if foetuses were present these were weighed, measured and (if small enough) preserved. Rarely in Sperm whales lumps of ambergris were found in the stomach, or sometimes discovered by treading on lumps lying amongst the coils of intestines and associated piles of tissue that I was wading through. All the samples were labelled and preserved in formalin and the position and time of capture recorded in the logbook.

These were the routine collections and observations made by all whaling biologists. Details of process and numbers and species of whales varied with time and location, but the work described above provided the raw material for the scientists back at NIO. It was a fascinating and exciting time. Locating, retrieving and sampling organs from an animal which could be up to 24 m long and although lying on its side stood some 2-3 m above the level of the platform was a huge experience for someone whose previous experience of mammalian anatomy was dissecting rabbits in the course of getting a degree!

The Discovery of a Turbulent Ocean

One of the principal reasons for setting up NIO was to give Britain a centre to study physical oceanography. In his research in the Southern Ocean, George Deacon established that the ocean currents, both shallow and deep, are part of a world-encircling system which carries heat, chemicals and food like a conveyer belt circling the earth. Jim Crease and John Gould write of the early attempts at NIO to understand the interactions between the atmosphere and oceans and to measure deep ocean currents. The ground-breaking work of John Swallow, in inventing the neutrally-buoyant float that could be tracked as it was carried by deep currents, showed how the deep ocean was not the placid place that we had assumed. Swallow's invention ultimately gave rise to the present-day Argo programme in which thousands of floats are tracked through the world's oceans and give vital data to test the ocean models and their contribution to understanding climate change. Steve Thorpe describes research on internal waves within the ocean and the important processes of mixing. Fred Culkin focusses on the critical role of salinity which, with temperature and pressure, determines the density of seawater crucial in driving the ocean circulation.

7

Ocean currents – entering the modern age

Jim Crease

Between the world wars, the accumulated efforts of many nations had described the average three-dimensional distributions of temperature, salinity, nutrients and dissolved oxygen of the oceans. Observational methods had changed little since the voyage of HMS *Challenger* in the 1870s. Temperature was measured using mercury-in-glass thermometers which, when turned upside down at depth, broke the mercury column and recorded the temperature to an accuracy of a few millidegrees even at pressures of 600 atmospheres. Chemical titrations to determine chlorinity (from which salinity was derived) and nutrients demanded great skill on board a constantly moving ship. Standard seawater of guaranteed salinity distributed from the International Council for the Exploration of the Sea (ICES) provided a calibration standard.

Long-term mean circulation of the oceans was inferred from the distribution of the properties (primarily temperature, salinity and dissolved oxygen) using the 'core method', developed by Wüst to interpret the measurements made on the 1926 *Meteor* survey of the South Atlantic. That survey showed such features as the cold water from the polar and sub-polar regions filling the deep ocean basins and Mediterranean water from the Straits of Gibraltar forming a high salinity layer at around 1 km and spreading across the North Atlantic, but gave no indication of the rate of flow.

The only progress towards that objective had been the development of the dynamical method by Scandinavian oceanographers, Helland-Hansen, Ekman, Nansen and Bjerknes. Just as atmospheric pressure measurements are used to calculate the strength and direction of winds at various heights, so the vertical and horizontal variations in temperature and salinity can be used to calculate pressure gradients in the ocean and hence the changes in current strength with increasing depth under the influence of the earth's rotation. However, the dynamical method had limitations. Firstly, throughout much of the ocean the horizontal changes in water properties are so small that the method can only be used over large distances. This limits the level of detail that can be obtained. Secondly, the method cannot be used near the equator where the Coriolis effect of the earth's rotation becomes zero. The most serious limitation was that the dynamical method could only determine the relative changes in current with depth. To determine currents absolutely an assumption had to be made that at some depth the currents were zero or some other predetermined value. Determining this 'level of no motion' became a holy grail for those studying ocean circulation. Oceanographic knowledge at the start of WWII was summarised in *The Oceans: Their Physics, Chemistry, and General Biology.*[1] The book was such an important landmark that the first copy to arrive in the UK was dispatched in the diplomatic bags from the Washington Embassy.

Learning about the interactions of atmosphere and ocean and the problem of measuring ocean circulation were early foci for the new NIO's physicists. In later years, Dr Deacon observed that he had formed his group for this endeavour from those who were surplus to requirements at other laboratories. I more or less fell into that category having joined the Royal Navy at age 13 and been invalided out at the tender age of 17. With a degree in mathematics from Cambridge, I was the last scientist to join the group at Teddington in 1951. This presented me with the opportunity of a career related to the sea that almost certainly resulted in more sea-going than if I had remained in the Navy!

New challenges

A main thrust of the work when I joined was still on surface waves. However, from the earliest days a related programme developed to understand the details of momentum transfer at the sea surface. Henry Charnock was recruited from Sir David Brunt's meteorology group at Imperial College to lead this work which he continued throughout his distinguished career.

Lough Neagh in Northern Ireland was the scene of pioneering work to measure wind stress over water, in collaboration with the Imperial College group and in particular with John Francis. Norman Smith recounts that it was in a pub on the shore of the Lough that he introduced Henry Charnock to Dr Deacon. Later, when Francis was Professor of Civil Engineering at Imperial College, Charnock and Francis attempted a small-scale laboratory simulation of airflow over water but the laboratory test needed a fluid with much higher viscosity than water. The primary problem was solved by persuading Tate and Lyle to donate about a ton of golden syrup, but there were other problems due to its hygroscopic properties. This left them with the 'sticky' challenge of how to dispose of the tank full of syrup!

The main observations at Lough Neagh were of wind speed and direction from a moored raft and water level measurements along the shore. They revealed a problem; the seiching (low-frequency 'sloshing') of the Lough made the interpretation of water level changes difficult. The Lough's remoteness from southeast England was also a hindrance and Charnock therefore developed a long-term project on the west London reservoirs of the Metropolitan Water Board near what is now Heathrow Airport. A permanent installation was established with anemometers and wave recorders mounted on an offshore mast and linked by cable to a hut on the bank. An experiment to measure the effect of surface films (like oil slicks) must have seemed bizarre to onlookers when they saw scientists walking along the reservoir edge apparently watering the surface using garden watering cans!

Ron Burling was a New Zealander who came to work with Charnock as a PhD student attached to Imperial College. He was one of the first to work on the shape of the energy spectrum of water waves. His 1955 thesis first reported an f^{-5} dependence of the spectrum at high frequencies.[2] Burling went on to a career at the University of British Columbia in Canada where he joined members of Professor Bob Stewart's group working on similar topics.

In the mid 1950s, Charnock's interest in the temperature and density microstructure in the oceans' surface layer led to a collaboration with the Royal Navy trying to understand the disturbance of the near-surface thermal structure that might result from the passage of a submerged submarine. Classified experiments were conducted near Malta using extremely delicate resistance thermometers (designed and built by Norman Smith) to determine the stratification. These were towed ahead of the bow disturbance of a small naval trawler whilst a submarine cruised in and below the well-mixed near surface layer and an aircraft from the Radar Research Establishment in Malvern carrying radiometers flew overhead. The scientific basis of this work was continued by John Woods in the same area in his 1970s work on the microstructure of the surface layer of the ocean where he photographed dye injected into the sharp density steps. John later held Chairs of Oceanography at Southampton, Kiel, and Imperial College and was Director of Marine Science for NERC from 1986 to 1994.

When still associated with Imperial College, Charnock had participated in an experiment in the Scilly Isles to investigate the vertical wind structure and the aerodynamic roughness of the sea surface. Working in a single location, they had to assume that horizontal temperature gradients had negligible effect. This proved not to be true and it was concluded that a successful experiment was "hardly possible in the barotropic westerlies" where there is only a small variation of wind velocity with height. So, in 1953, an expedition was organised to Anegada in the Caribbean in the more uniform trade-wind zone. The drag coefficient calculated from that experiment is surprisingly close to modern estimates.[3]

Charnock's formula relating drag to wind speed and surface roughness stemmed from the data from the London reservoirs. The classical neutral stability vertical profile of wind speed in the boundary layer has a logarithmic variation with height, and a scaling length related to the height of the waves and ripples. The dynamic nature of the sea surface, unlike the state of affairs at a fixed location over land, leads to a scaling length that is dynamically related to the surface wind speed. Actually, Charnock was somewhat embarrassed that the formula was attributed to him and said that it had been developed jointly with Tom Ellison of Cambridge University. (The Charnock parameter is one of two internationally recognised parameters to which members of NIO can lay claim, the other being the Thorpe parameter.) It was with Ellison, in 1957, that Charnock investigated fluctuations of temperature and humidity in marine air by making measurements at the top of Blackpool Tower.

Deep ocean currents

In 1953 we all – physical oceanographers, biologists, geophysicists and engineers – came together at Wormley. During the war the building had been the workplace of the eminent cosmologists Herman Bondi, Tommy Gold and Fred Hoyle working on the development of centimetric radar.

If we were to make progress in understanding the ocean circulation some means had to be devised to make direct measurements of currents well below the surface of the open ocean. Drogues (small surface buoys attached to large subsurface parachutes) could be tracked as they drifted with the currents, but with only celestial navigation to determine their positions the tracks were uncertain except when measured over several days. There were current-measuring instruments (current meters) available but all had major deficiencies and there were few measurements in the deep ocean off the continental shelf. In one of the first publications from NIO Ken Bowden summarised all these measurements.[4] They fitted into a single table and none of the experiments lasted more than a few days. Clearly, neither these nor drogue measurements would suffice.

Meteorologists had no such problem. Anemometers measured wind at the surface and at hundreds of radiosonde stations balloons measured the temperature profile every 6 hours. The balloons' horizontal displacement as they rose (tracked by theodolite) gave the winds at various levels. Atmosphere sampling was well defined; the horizontal scale of depressions and anticyclones is typically of the order of 1,000 km and could be mapped by measurements hundreds of kilometres apart.

Dominant horizontal scales in the ocean were only known from the changes of temperature and salinity seen across ocean basins and where sharp gradients (fronts) occurred across major currents. The timescale for an atmospheric depression to pass was measured in days. The sparse ocean measurements were insufficient to define any timescale of variability other than seasonal and diurnal variability in the upper ocean.

Tom Tucker, Frank Pierce and I discussed how we might measure currents in the open ocean. Using the radiosonde analogy, I suggested having free-falling sound-emitting devices tracked by an array of acoustic receivers, thereby measuring the horizontal displacement as the device sank. If this succeeded, then we could measure the velocity field from top to bottom uncontaminated by instrument errors. Although this would be a huge advance it would still only give a 'snapshot' and would not reveal the time-dependence of the circulation.

There was a trial of this method on *Discovery II* in October 1954. To slow the descent, the device was hung below parachutes attached by piano wire. We all lined up around the ship's rail, parachutes in hand, and slowly moved in step towards the stern as the device was paid out. The trial showed that we had been over-optimistic about the efficiency of the acoustic tracking method, but the point is that ideas, even those put forward by the youngest staff, were allowed to be pursued and tested.

John Swallow and his float

In summer 1954, Dr Deacon invited Professor Edward (Teddy) Bullard from the Department of Geodesy and Geophysics at Cambridge to visit NIO. He was accompanied by John Swallow; a PhD student. A principal topic of conversation during the visit was the art of deep-sea instrumentation in which the Cambridge Department was building an enviable record. Bullard was carrying out pioneering work on the heat flow through the ocean floor. Swallow's PhD was on seismic experiments he had made during the global voyage of HMS Challenger (1950-52) to examine the structure of the ocean floor.[5] Tom Gaskell, the scientific leader, went on to become a senior scientist with British Petroleum and Steve Ritchie, Challenger's commanding officer, later became Hydrographer of the Navy. Deacon invited Swallow to join the staff of NIO.

When Swallow thought about the sinking parachute experiment, he realised that the system was impractical, but felt that some other acoustic tracking method might hold promise. What was needed was a device that would remain at a fixed depth and so could be tracked for an extended period. Such a method would have the enormous benefit of addressing the issue of time-dependence of the circulation, provided one could locate the float for long enough. In this, his thoughts were shaped by his experiences during his PhD. Through his seismic work on HMS *Challenger* he was acutely aware that the velocity of sound in

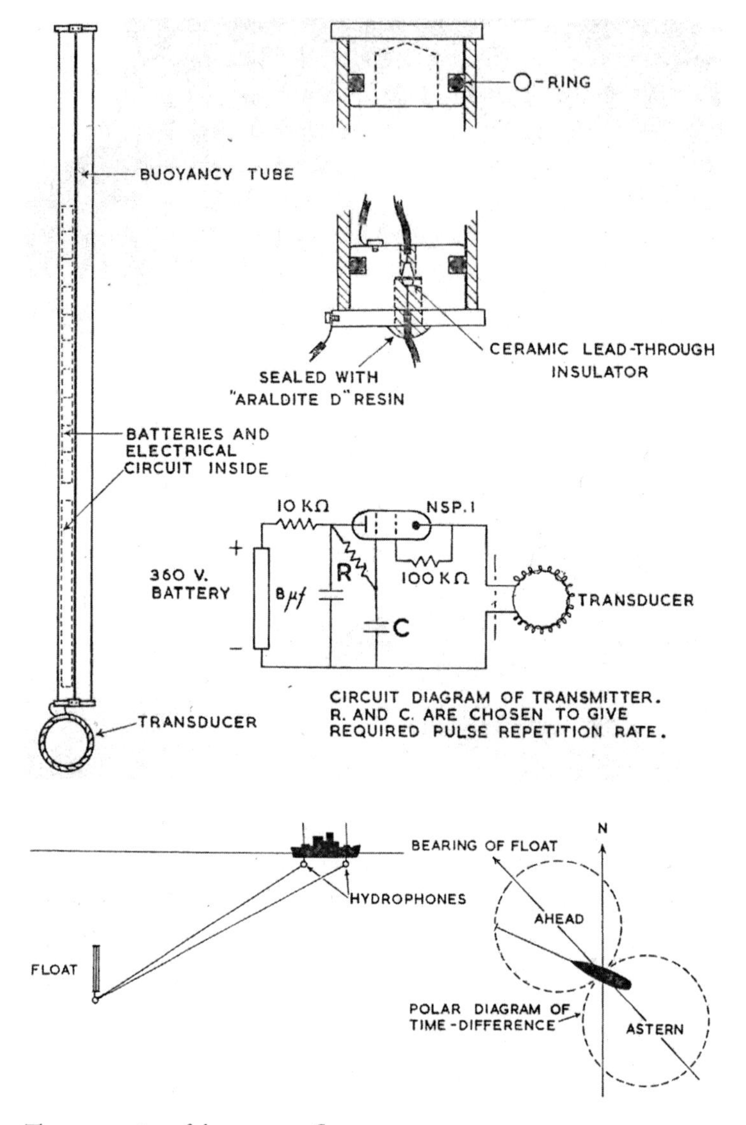

Top: *The construction of the prototype float.*
Bottom: *The method of tracking floats from Swallow's 1955 paper.*[6]

solid rock was much higher than that in water. That, combined with the higher density of the rock, meant that water was an order of magnitude more compressible than most solids.

It occurred to him that it might be possible to make an instrument less compressible than seawater and yet with enough buoyancy to give a useful payload. Such a device, if suitably weighted, would gain buoyancy as it sank and the ambient pressure increased (1 atmosphere for every 10 m). At some depth it could have a density equal to that of the surrounding water.

John Swallow with an early float on RRS Discovery II *watched by (left to right) the ship's Donkeyman (Charlie Fry), a greaser and the ship's cat.*

In building such a neutral-buoyancy float, soon to be known as a Swallow float, NIO's continuing links to the Royal Navy stood us in good stead. Acoustic techniques were commonplace in the lab and previous work at 10 kHz frequencies (giving communication out to a few kilometres) meant that suitable transducers were readily available.

Swallow had to find materials with appropriate dimensions for the pressure case, balancing density and compressibility. At the time, he had taken over my place in a bachelor flat in Kingwood, just down the road from NIO. (I had become disqualified from residence by marriage.) The other inhabitants, Tom Tucker, Roland Cox, Arthur Stride and Tony Laughton remember John spending long hours using his Curta mechanical calculator working out what would be possible. He came up with the surprising answer that standard aluminium scaffold poles would be almost ideal. The construction of the prototypes was elegantly simple, though painstaking in practice. To get enough buoyancy to support the transducer, batteries and circuitry the wall thickness of the tube had to be reduced slightly. The modification of the 3 m poles was tricky. They were placed in a wooden trough filled with caustic soda that fizzed away like a witches' brew in the yard at NIO until the correct value was achieved. The float end caps were sealed with, then quite new, O-rings.

Swallow determined the weight of each float in water of known density, so that it would stabilise at the intended depth. The balancing was done in a 6 m-high cylinder of thoroughly-stirred salty water in the stairwell of NIO. Internal weights were added to make the float slightly negatively buoyant and its weight in water measured using a simple chemical balance from which the float was suspended. The water density was determined using specific gravity bottles. Typically the addition of 1 g to a 20 kg float would make a 10 m difference in float depth.

By May 1955 a float designed to operate at 2 km depth was successfully tested on the RV *Sarsia*. The electronics circuit had a single electronic vacuum tube and a capacitor discharging into the transducer.[6]

Following this success, in 1956 multiple float deployments were made in the eastern North Atlantic off Portugal above the 4 km-deep Iberian Abyssal Plain. The challenge there was navigation. The area was out of range of both the European Decca and the US Loran-A radio navigation systems and so position finding depended on astronomical sights and dead reckoning, techniques unchanged since the introduction of Harrison's marine chronometer in the 1770s. Celestial navigation alone was not sufficiently accurate to determine unambiguously the movement of a slow-moving float. However, thanks to developments in precision echo sounding by Tom Tucker and his colleagues, Swallow could use the small hills, a few tens of metres high, that stuck up out of the flat abyssal plain as reference points by which to locate moored buoys carrying radar reflectors. These were used to fix the ship's position while listening with hydrophones to the 10 kHz sound pulses (pings) from the floats. Hydrophones fore and aft were used to determine the bearing of the float and then to fix its horizontal position (and estimate its depth) by triangulation. In the space of a month, more measurements were made of deep-ocean currents than ever made before, tidal oscillations were revealed along with a small meandering 2 cm/s drift at 2,000 m depth. Those were heady days.

Working with Woods Hole

It was about this time that John Swallow started a life-long friendship and collaboration with Henry (Hank) Stommel of the Woods Hole Oceanographic Institution (WHOI) in Massachusetts. Stommel had come up with the idea of a neutrally buoyant float at the same time as John but completely independently.[7] He envisaged floats being tracked at long range by means of 'bombs' exploded in the SOFAR (Sound Fixing And Ranging) channel (the minimum in the profile of the speed of sound in sea water, usually at about 1,000 m, that allows sound to travel long distances). It was over 20 years before Stommel's dream of long-range tracking was realised.

A very significant development had taken place in our understanding of why strong poleward currents such as the Gulf Stream in the North

Atlantic are found along the western boundaries of oceans. A dynamical theory of this 'westward intensification' had first been proposed by Walter Munk at the Scripps Institution of Oceanography and finally resulted in a simplified exposition by Stommel in a paper that attracted much attention.[8] The theory was supported by laboratory experiments using a pie-shaped 'ocean' on a rotating turntable. The theoretical approach was complemented by measurements of the temperature/salinity (hydrographic) structure of the Gulf Stream system made by Fritz Fuglister, Val Worthington and their WHOI colleagues.

The close collaboration between NIO and WHOI was probably aided by the fact that the style of work at WHOI was similar to our own. Our bosses (Columbus Iselin and George Deacon) were not dissimilar. Each had a profound feeling for oceanography and each relied on his personal judgment of potential recruits. Fuglister (leader of WHOI's hydrographers–'water catchers') had been spotted as a future oceanographer by Iselin, when he found him painting murals on the wall of a Cape Cod church during the Depression. Stommel, Fuglister and Worthington were members of the unofficial Society of Subprofessional Oceanographers, SOSO, sharing the distinction of having no professional oceanographic qualifications! (An excellent description of these events from a US perspective can be found in the introduction to the book *Evolution of Physical Oceanography*.)[9]

In early 1957, Stommel and Worthington planned a survey of the Gulf Stream, using WHOI's RV *Atlantis* (a beautiful 142 ft steel-hulled ketch built in Copenhagen in 1930). Further theoretical work by Stommel on global-scale ocean circulation predicted that there should be a deep countercurrent under the Gulf Stream flowing southwards down the western boundary of the Atlantic.[10] Whilst Worthington was unconvinced, Swallow agreed to bring *Discovery II* to the USA and to use his floats to explore the currents below the Stream. They did, indeed, discover the deep undercurrent[11] but the four-week experiment was too short to be certain whether this was a rare and anomalous event or the even rarer case of a theoretical prediction preceding a confirmatory observation.

In 1957 and 1958 during the International Geophysical Year (IGY), again with scientists from WHOI, *Discovery II* made hydrographic sections (principally transatlantic ones on 24°, 36° and 48°N) as a major contribution to the first full-scale physical survey of the Atlantic since the 1920s *Meteor* cruises. Only the best carefully calibrated thermometers had the precision needed in deep water. We were fortunate in having very stable thermometers that had been 'acquired' in Germany at the end of the war as a result of the efforts of J.N. 'Jack' Carruthers, who was then oceanographer in the Hydrographic Department of the Navy. He went to Germany at the end of hostilities and worked with German colleagues to ensure the survival and re-establishment of the major German

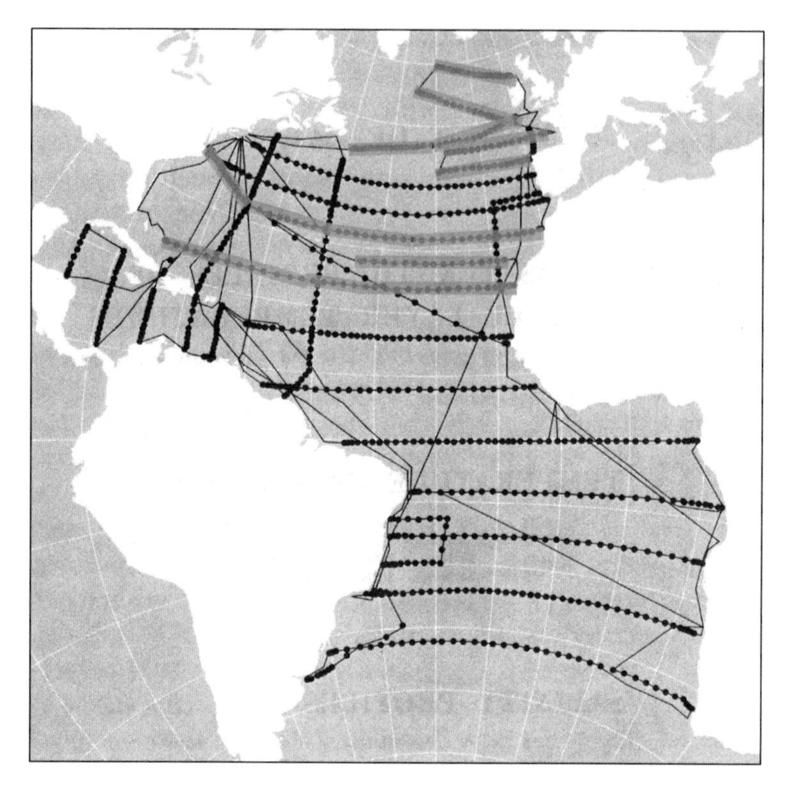

The IGY station positions. Those worked by Discovery II *in the North Atlantic are highlighted.*

oceanographic labs. Later, he joined NIO with his colleague Cdr Lawford and brought with him, in addition to the thermometers, the Hydrographic Office's oceanographic library and his own personal book collection. These, with the Discovery Investigations library, were the foundation of the NIO library.

The results of the IGY survey were published by WHOI in an atlas by Fuglister and Worthington that remained a definitive reference source for many years. Surprisingly there were no NIO publications based on the IGY sections but their true value was realised much later when repeats of the IGY 24°N section in the 1980s and 90s allowed changes in the ocean and its poleward transport of heat to be determined.[12]

Following the use of Swallow floats in the northeast Atlantic and under the Gulf Stream the time had come to venture further into the ocean and to explore its deep circulation. Stommel and Swallow planned a year-long experiment for 1959-60 in the western north Atlantic to measure the currents at and below 2 km. This would test Stommel's hypothesis that, away from the deep Gulf Stream undercurrent, there would be a sluggish interior flow. Stommel obtained funding for the experiment from the US National Science Foundation.

Left: *Jim Crease handling a hydrophone aboard* Aries. *Floats are stacked against the ship's rail.*
Right: *The Aries in full sail. (By courtesy of Woods Hole Oceanographic Institution Archives.)*

Fortuitously, the American tobacco millionaire, R.J. Reynolds, had recently donated the 115-ft ketch, *Aries*, to WHOI, complete with crew. WHOI offered the use of *Aries* to NIO, and specifically to Swallow, for 18 months as a platform for the work. (*Aries* had been built by Camper and Nicholson's in Gosport in 1952 and still sails the oceans as the historic yacht *White Heather*.) The original crew, who were used to a life spent mostly alongside in the Bahamas, didn't last long and most were replaced by a WHOI crew used to the whims and working habits of scientists. The ship was based at the Bermuda Biological Station known as the Bio Station.

The scientific party were Swallow and myself from NIO, accompanied by various WHOI people (particularly Gordon Volkmann). Others included Charnock and Allan Robinson (then a student and eventually Gordon McKay Professor of Geophysical Fluid Dynamics at Harvard). The routine was that we sailed west for a couple of days to reach the area of the experiment and for the next eight days or so we would launch and track up to half a dozen floats typically at 2,000 and 4,500 m depth over a 100 km-square region in the Sargasso Sea. Between May 1959 and August 1960, *Aries* made 26 such cruises.

Tracking the floats was through triangulation using hydrophones fore and aft. The position of the *Aries* was determined with a Loran-C set built by Bob Walden of WHOI, there being no commercial set available. In between float tracking we made many water-bottle station observations. Back in the Bio Station there was much to do over the weekend before the next cruise–preparing new floats and analysing our salinity samples to the third decimal in salinity (thanks to Roland Cox's accurate and robust NIO Thermostat Salinometer which replaced titrations). For the remaining hours John and I had the pleasure of the company of our young families who were living at the Bio Station.

We had expected the floats to move slowly and to stay within a small region and be tracked for months. They had therefore been fitted with clocks that switched on the acoustics for just an hour or so each day to conserve the batteries. In practice, the currents were much faster and more variable than expected and the clocks had to be abandoned.

After 15 months work at sea, with 70 or so floats tracked for periods up to a couple of weeks, we had for the first time a clear picture of an energetic deep ocean with current speeds of up to 20 cm/s at 4,500 m. What we had discovered was that the open ocean dynamics were dominated by 'weather' rather than 'climate' and had a small horizontal scale. Near Bermuda this scale was around 50 km (compared to the equivalent scale of 1,000 km in the atmosphere). Despite the revolutionary nature of these findings publication of the results was slow,[13,14] but word of what we had discovered got around and was to set the scene for the exploration of what became known as ocean mesoscale variability.

The Atlantic and beyond

Closer to home, the area from the Shetland Islands to Greenland was of great interest to our colleagues in the Fisheries Laboratories at Lowestoft and Aberdeen. It is a region where cold, dense water (below 0°C) formed in the Arctic enters the North Atlantic through the Færoe-Shetland channel (the deepest connection between the Norwegian Sea and the Atlantic) and the Denmark Strait (between Iceland and Greenland). Much of our collaboration with the fisheries scientists was in surveys of the interesting hydrographic and chemical structure in the area; routine in nature but rewarding in insight and in the case of the Færoe-Shetland Channel building on the foundation of some of the earliest repeated measurements started in 1927.

Despite our knowledge of the hydrography there were no measurements of the intense near-bottom flow of Arctic water. So we launched Swallow floats near the entrance to the Færoe Bank Channel and placed the floats close to the bottom by hanging a rod weighing a few tens of grams below the float on a 20 m length of thin wire and making the entire instrument slightly negatively buoyant at the target depth. As soon as a cm or so of

John Swallow (L) and Gordon Volkmann making final adjustments to a neutrally buoyant float aboard the MV Erika Dan *in 1962. The circular object is the magnetostrictive nickel scroll sound transducer. (By courtesy of Woods Hole Oceanographic Institution Archives.)*

the rod touched bottom the excess negative buoyancy was removed and the float stabilised. In the strong currents (around 1 m/s) that we measured, the float could follow the bottom-hugging flow with minimal drag. We were thus able to make the first direct observations of the overflow from the Norwegian Sea that had previously only been deduced from the hydrographic data.[15] John Swallow also used his floats aboard the *Erika Dan* in 1962 to measure the deep outflow from the Norwegian Sea through the Denmark Strait on a cruise led by Val Worthington from WHOI.

The International Indian Ocean Expedition (IIOE) in 1959-65 was a co-ordinated programme of the newly-established Intergovernmental

John Swallow fitting an NIO water bottle to the wire during a hydrographic station on RRS Discovery *during the Indian Ocean Expedition.*

Oceanographic Commission (IOC) of UNESCO. Participation required a major commitment from all groups in NIO. NIO work focused on the Arabian Sea in collaboration with our colleagues, Stommel, Bruce Warren and John Bruce, at WHOI and was carried out on the new *Discovery* and the Scripps ship the *Argo*.

Hydrographic surveys were carried out of what was then an oceanographically poorly known area strongly affected by the seasonal monsoons. In both monsoon seasons we occupied two meridional lines of stations from the Arabian coast to south of the equator, followed by a zigzag survey of the coastal zone from Mombassa up to the Horn of Africa and including the strong Somali current – another of the western boundary currents but one that, unlike the Gulf Stream, reversed direction seasonally with the monsoon winds.[16]

The presence of an eastward flowing undercurrent on the equator in the Pacific had been established by the work of Townshend Cromwell at Scripps in the early 1950s. The monsoonal regime of the Indian Ocean raised questions about the nature of such an undercurrent in that ocean. John Knauss and Bruce Taft from Scripps had shown that the undercurrent was barely perceptible in the NE monsoon in 1963. *Discovery*'s meridional sections in the two monsoons in 1964 used direct reading current meters (DRCMs) lowered from the ship. They measured the shear between the surface and 200 m (hence the movement of the ship was not so important) and showed that there was indeed an undercurrent and that it was strongly influenced by the monsoon.[17] These early results were a catalyst for a long collaboration between John Swallow, Henry Stommel and the German scientists Fritz Schott and Detlef Quadfasel (Institut für Meereskunde, Kiel and Hamburg) on the equatorial oceanography of the Indian Ocean. It lasted until Swallow's death in 1994.

We were all determined to extract the most from the resources that our new ship had to offer so it became normal to have equipment deployed from three winches at the same time. We used the fo'c's'le winch for hydrographic water bottle casts, the midships winch with its conducting cable for the DRCM current meters and the after hydro winch for running a series of plankton nets. (All these winches were powered by 1870 *Challenger*-era steam reciprocating engines!) Most of the time this worked well, but occasionally in the Somali current the strong shear (sometimes as much as 4 kts difference between the surface and 200 m) would result in tangled wires. The operations put tremendous demands on the ship's officers and crew, but they coped superbly and we were a happy ship.

The cruise also provided an opportunity to test two new pieces of equipment that were to become the workhorses of oceanography from the 1970s onwards. The first of these was the prototype of a current meter developed under a NATO contract at the Christian Michelson Institute in Bergen, Norway. It recorded its data internally on magnetic tape but also had an acoustic link and so could relay current speed and direction data back to the ship without the conducting wire needed for the DRCMs. It was later manufactured commercially as the Aanderaa RCM4. Neil Brown from WHOI also brought along the prototype of his Conductivity-Temperature-Depth (CTD) probe that transmitted its data through a conducting cable. This, and its successors, ultimately replaced our dependence on water bottles and thermometers.

Mapping currents was an important part of the IIOE work and we used a variety of techniques, all dependent on celestial navigation or, where possible, land bearings for our absolute positioning. One method exploited the electromagnetic effects of seawater (an electrical conductor) moving through the earth's magnetic field and generating an electric field. When underway we towed a cable with electrodes at each end, a device designed by Bill Von Arx at WHOI. The geomagnetic-electrokinetograph (GEK) was used to demonstrate that the strength of the Somali current could reach 6 kts at the height of the SW Monsoon. We spent an immense amount of time in determining where we were.

The prototype Bergen current meter and below it the prototype conductivity-temperature-depth (CTD) instrument deployed from Discovery*'s midships winch during the IIOE. The scientist reaching over the side is Bruce Hamon, a visitor to NIO from CSIRO, Cronulla, Australia.*

A consequence of the geostrophic balance (between the density/pressure field and the earth's rotation) was that the density field in the Indian Ocean sloped strongly up towards the coast, enough indeed to bring sub-thermocline cold water to the surface, resulting in both the mass extinction of the tropical fish and the diversion of the boundary current from the coast onto an eastward path across the Arabian Sea. This upwelled deep water on the surface off the Somali coast in August 1964 was cooler than any surface water we encountered on the ship's passage back to Plymouth.

Final memories of IIOE were of the hydrographic stations we made in the Red Sea on our way home in September 1964. *Atlantis II* and *Albatross* had earlier reported anomalously high bottom temperatures (ca. 25°C) in the area. We stopped for a few hours and it was with great excitement that temperatures of 44°C were found in what was essentially a pool of saturated brine full of salts emerging from the seafloor with a salinity of 270‰ (normal oceanic values are around 35‰). The boundary between the brine and the overlying water could even be seen on the echosounder. The water spattered from the sampling bottle onto the hot deck quickly turned to a pile of salt. This was the start of an era of discovery of extremely hot vents in many mid-ocean ridge systems, though our lab did not follow this up to any major extent at the time.[18]

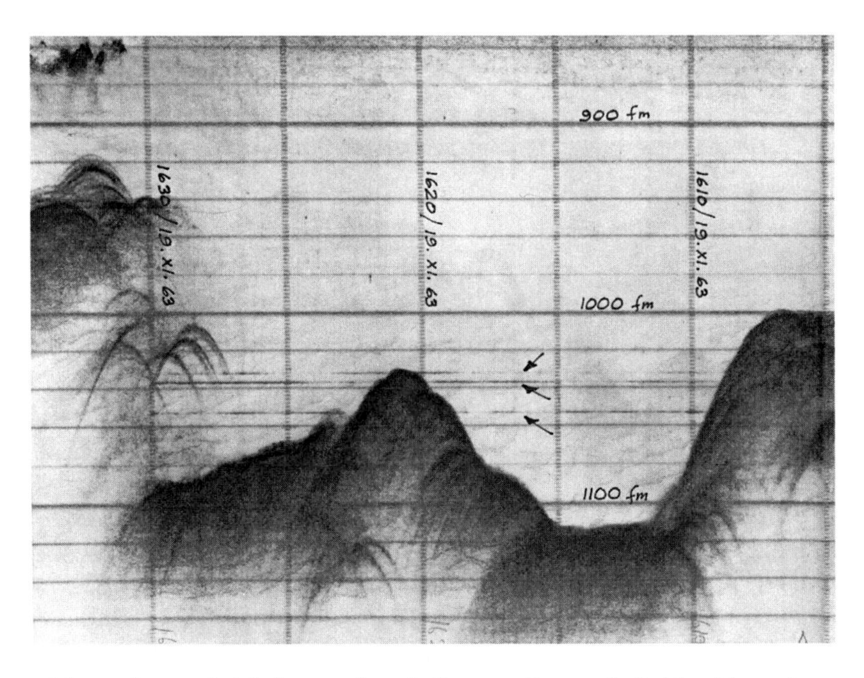

Echosounder record of the brine pools in the Discovery Deep in the Red Sea. The top layer of the hot brine shows up as horizontal lines (arrowed).

Swallow's development of the float revolutionised our ability to observe the deep ocean. The measurements from the *Aries* had shown that deep-sea currents were much stronger and more variable than expected from dynamical calculations. The study of that variability was to be the focus of NIO's work on ocean circulation from the mid-1960s to the end of the period covered by this book and would require new instruments to be developed.

8

Exploring ocean variability

John Gould

My role in this story started when I joined *Discovery* Cruise 10 in Plymouth in February 1966 as a student helper. I was in the second year of my PhD at the University College of North Wales in Menai Bridge supervised by Professor (Jack) Darbyshire. Cruise 10, led by John Swallow, was my first taste of deep-sea oceanography and, after having previously worked in the Irish Sea where the depths were a few tens of meters and stations took only 10 minutes, I remember it came as shock to work in 5 km of water where they took several hours.

Experiments with current meters

Cruise 10's main objective was to measure the 'patchiness' of the Mediterranean water near Madeira. (Warm salty water from the Mediterranean Sea flows out of the Straits of Gibraltar and spreads in the Atlantic at a depth of about 1 km. Its high salinity can be detected as far away as the Caribbean.) It was also an opportunity to test new equipment. I was responsible for analysing data from current meters that recorded their data internally and were to be deployed on moorings.

John Swallow had first worked with moored current meters aboard the *Erika Dan* in January-April 1962 on a cruise led by Val Worthington from WHOI. That cruise, a follow-up to the work during the IGY in 1957-58 on *Discovery II*, occupied hydrographic sections around the southern and eastern coasts of Greenland to map the deep overflow of cold, fresh water from the Denmark Strait and Færoe Bank Channel. Swallow was there to measure the deep currents using neutrally buoyant floats. The moorings on the *Erika Dan* cruise were supported by surface buoys and carried current meters designed by Bill Richardson of WHOI and positioned in the deep overflow. In fact the current meter experiment produced very little data and showed just how difficult it was to make such measurements. As Gerry Metcalf of Woods Hole commented, "*This modern electronic oceanography may be the wave of the future, but at present it is hideously expensive and damned unproductive*".

On cruise 10 we used three current meter designs; two from the USA (Braincon and Geodyne, the latter being a derivative from the Richardson instruments) and one from the UK. This was a prototype of a current meter later produced by Plessey. It was similar to the NATO meter tested on the IIOE but used an impeller in place of the Savonius rotor. The US current meters recorded their data on 16 mm ciné film, the Plessey on ¼ in magnetic tape. The heart of the Plessey (and NATO) current meters was an electromechanical 10 bit binary encoder.

The moorings on cruise 10 were more sophisticated than those used on *Erika Dan*. They had subsurface buoyancy to isolate the instruments from wave action (and from the attention of passing ships and fishermen) and relied for their recovery on an acoustic release that, when commanded from the ship, would fire an explosive bolt and separate the mooring from its anchor. The releases were developed from the NIO 10 kHz control system used to open and close biological nets. NIO first used acoustic releases in the Færoe Bank Channel in 1965, only one year after WHOI had used their first (commercially manufactured) ones. Cruise 10 was our first attempt to use them in the deep ocean.

Deploying moorings was hazardous. The wire lengths and their connecting shackles were wound under tension onto the main trawl winch with the shackles wrapped with canvas to stop them snagging on the wire as it unwound during deployment. The anchor (half a ton of scrap chain) was then attached to the mooring wires and lowered over the side. First the acoustic release was attached (the ship's radar was switched off in case its radiation should fire the explosive bolt), then the current meters and finally the buoyancy was added. At this stage, with luck and good judgment, the ship would be in the right water depth so that the instruments would be at their intended depths and, more importantly, the buoyancy would neither remain at the surface nor go so deep as to implode. If the depth was not right then the mooring had to towed slowly until the depth was correct and was then cut loose to sink out of sight. Dick Burt, the ship's netman, was the key person in rigging the moorings. It is remarkable that he survived to reach retirement without serious injury caused by a wire breaking.

We recovered the first mooring after a three-day deployment and I read the data from the Braincon current meter. The other two current meters had data recording problems. During the second deployment, the wire snagged and the entire rig was lost. So the whole episode was almost as unsuccessful as Worthington's Greenland measurements had been.

Also on board was 70-year-old J.N. Carruthers testing one of his simple low-cost current measuring instruments. The 'abyssal pisa' had a

J.N. Carruthers holds court on the foredeck of RRS Discovery *Cruise 10. His attentive listeners are David Pugh, then a PhD student at Cambridge and who went on to be President of the IOC, and Prof Bob Smith (in checked shirt) from Oregon State University. Driving the winch in the background is 'Dicky' Dobson.*

tube filled with hot water containing a baby's feeding bottle. The bottle was part-filled with table jelly on the surface of which floated a magnet. The remainder of the bottle was filled with oil. The principle was that the device would be lowered to the seabed where it would be deflected from the vertical by the currents and would remain there until the water cooled, the jelly set and fixed the magnet in place, so recording the strength, (angle of the set jelly) and direction (orientation of the maximum slope to magnetic north). It too, collected rather little information. Other current measurements were made by standard 10 kHz Swallow floats. These were now tracked by towing two pairs of hydrophones behind the ship and detecting whether the float was ahead or astern, to port or starboard.

This, my first *Discovery* cruise, despite awful weather in the Bay of Biscay and the sad death of one of the ship's stewards on the last night of our port call in Madeira, had whetted my appetite for 'proper' oceanography. At the end of the cruise, Swallow came to find me on Plymouth station and said "*If you'd like to work at NIO when your PhD is finished, just let me know*". I joined the NIO as a scientific officer in September 1967.

In January 1967, Swallow again joined Worthington, this time aboard the Canadian Research Ship *Hudson,* in a second attempt to collect long

Mooring deployment. Note the lack of safety equipment.
Top Left: *The acoustic release. (Dick Burt, kneeling, and John Swallow, standing.)*
Top Right: *An RCM4 current meter. John Cherriman, the leader of the NIO mooring team (upper centre).*
Below: *A 4 ft diameter buoyancy sphere.*

observations of the flow of deep water from the Denmark Strait and to measure the winter hydrographic conditions in the Irminger Sea. The cruise lasted over three months; it was stormy and cold and in some respects another disaster. While the Swallow float results and the hydrography data were immensely valuable, very few of the WHOI moorings were recovered and of those only one current meter out of 30 returned a useable record. Worthington remarked that, since the measurements were funded by the US Office of Naval Research (ONR) then at least the money would not be available for use in the Vietnam war![1,2]

I worked with Swallow to develop NIO's ability to use current meters. The Aanderaa RCM4 (the commercially produced version of the NATO current meter) proved much more reliable than its Plessey 'cousin' and became our workhorse.

We learned a great deal about moorings through our close links with WHOI, helped greatly by an extended visit to NIO in 1970 by Ferris Webster who, together with Nick Fofonoff, led the scientific work of WHOI's 'buoy group'.

WHOI had started to maintain moorings at 'Site D' on the continental slope south of Woods Hole and NIO started a similar site on the Meriadzek Terrace, a plateau with depths of around 2,000 m on the continental slope southwest of Lands End. Our aim was to be able to deploy a subsurface mooring for a month (the period in which the RCM4 filled its data tape with a sample every 10 minutes). This was not easy initially and many were lost through corrosion (even of stainless steel), through being damaged by long-line tuna fishing and when their aluminium 'sausage'-shaped floats were dragged too deep by the strong currents and imploded.

Plastic-coated wire and 48 in diameter steel spheres that provided lower drag and greater buoyancy were major advances. We collaborated with the Vickers company who put samples of non-metallic mooring lines and syntactic foam (a matrix containing tiny glass spheres) on our test moorings. We used a wide variety of ships to maintain the moorings and Dennis Gaunt designed a clever self-contained double-barrelled winch that could be moved from ship to ship. It kept the mooring line under low tension at the inboard end while the outboard end held the full mooring weight and this eliminated the chance of the wire snagging on shackles during deployment. The idea for the winch came from the textile industry. Mooring operations were also made safer by replacing the explosive bolts with plastic (non load-bearing) pyros. Should they fire prematurely, anyone nearby might be burned but not killed.

My lasting memories of that era are of the long hours spent searching for 'missing' moorings. I wish I had a photograph of Gaunt, in his

John Cherriman preparing a batch of Aanderaa current meter recording units.

grubby duffel coat, hunched over the Mufax recorder searching for faint signs that the acoustic release had responded, and alongside him a tin lid full of cigarette ends and ash; never a pretty sight in the small hours of the morning.

By the end of the 1960s, we could collect month-long records reliably and the data were scientifically valuable. We confirmed the existence of a steady poleward-flowing boundary current along the Atlantic's eastern margin and studied the tides within the ocean.[3,4] The Aanderaa current meters had temperature sensors (US instruments did not) and so we were able to study internal waves. Work on my PhD continued while I was at NIO, but I changed its focus to a study of moored current meter data. My results were used by Chris Garrett and Walter Munk to confirm their theories on a universal internal wave spectrum.[5]

At sea with the Russians

In the 1960s and early 70s, there were so many current meter and mooring designs that it became important to see whether all the measurements were consistent with one another. In 1970, John Swallow became chairman of Working Group 21 of the Scientific Committee for Oceanic Research (SCOR) charged with carrying out current meter and mooring inter-comparison experiments. The first at Woods Hole Site D involved US, German, UK and Norwegian instruments on a pair of moorings, one with surface buoyancy and the other with subsurface. The records showed that instruments recorded higher current speeds on moorings with surface buoyancy and a second experiment was planned to determine the cause.

The experiment was to take place aboard the Soviet RV *Akademik Kurchatov* in early 1970 but since Swallow was already committed to leading a *Discovery* cruise to the Mediterranean (MEDOC '70) he asked me to take his place. The *Kurchatov* participants were to be Bob Heinmiller, head of the WHOI Buoy Group, and scientists from Bedford Institute (BIO) in Canada, Institut für Meereskunde (IfM), Kiel, West Germany and from the IfM in Warnemünde in East Germany, and of course the Russians from the P.P. Shirshov Institute of Oceanology in Moscow. (There had been no contact between the labs in Kiel and Warnemunde, only 120 km apart, since the closing of the border between east and west Germany in 1952.) The 'western' scientists would join the ship in Dover in February as it passed *en route* from Kaliningrad to the subtropical NE Atlantic. The end port was unknown, as was the duration of the cruise.

This was the era of cold war secrecy and communication with Moscow was via telex messages and occasional telephone calls that repeatedly told us of delays. But as March approached, departure appeared imminent

John Gould (L) preparing to attach a current meter to the mooring wire on RV Kurchatov. *Kyrill Chekotillo is assisting.*

and we and our current meters headed to Dover to await the *Kurchatov's* arrival. We embarked from a pilot cutter and headed off towards an unknown destination.

Life on board seemed strange. We were privileged visitors and ate with the officers and senior scientists while most of our Russian colleagues ate in the crew mess (an interesting perspective on communism). There were constant announcements over the public address system – Bob Heinmiller became so annoyed that he shook his fist and swore at the loudspeaker in our lab. A Soviet colleague said "*Don't worry Bob, it can't hear you*" which made him think that perhaps it could. There were rendezvous at sea with other Soviet ships for the transfer of people and equipment and an excuse for a party. Due to secrecy, we were not allowed to see the echosounder or to send messages about the work.

The SCOR moorings were embedded in what we discovered, much later, was a huge array of moorings near the Cape Verde Islands, POLYGON-70, that was being maintained for six months to map the kind of mesoscale eddies that Swallow had discovered in 1960 on the *Aries*. After five weeks (we had only expected to be away for 2 or 3), we docked in Monrovia, Liberia (the only country in West Africa for which we did not have visas).

The subsequent analysis of the data was done that summer at WHOI. It gave further evidence that most current meters (including the Russian ones) gave spuriously high readings on surface moorings (as used in the

POLYGON array). International sensitivities forbad us to say that in the reports, so we spent many hours with the Soviet member of the WG, Kyrill Chekotillo, finding a suitable way of saying what was unacceptable to our Soviet partners. Later, during my post-doc stay in Woods Hole in 1972, during which I took part in the third and final WG21 experiment, I published a paper that defined the surface moorings biases.[6] It was not until 1992 that a young Russian scientist, Nikolai Maximenko, was able to publish an equivalent paper in a Russian journal.

As a postscript, during the Kurchatov cruise, I found that Bob Heinmiller had been visited by the CIA and the Canadians by the Mounties asking them to report back after the cruise. Someone from the Royal Navy turned up at NIO asking to speak to me a week after we sailed. We heard later Kyrill gave up science and became a 'diplomat'.

Air-sea interaction

By the late 1960s we had a much-improved arsenal of observational techniques. We could deploy current meters reliably, we had Swallow's floats and we had replaced our water bottles and thermometers with Conductivity/Temperature/Depth (CTD) profilers that gave a continuous record from top to bottom rather than the measurements at 20-30 discrete levels. This allowed us to study a wider range of phenomena and to return to the difficult topic of interactions between the atmosphere and ocean.

The most extreme of these interactions is the formation of cold dense water in Arctic and Antarctic winters; water that ultimately fills the oceans' deep basins by a process called deep convection. The temperature and salinity of the water that flows into the Atlantic through the Straits of Gibraltar, which we had studied on cruise 10 (and other subsequent cruises), are determined by wintertime cooling in the Western Mediterranean – a process similar to the polar deep convection but in a much more accessible region. In 1969 and 1970, NIO scientists led by Swallow and Charnock took part in an investigation of the deep convection process in an area south of Toulon where the ocean is mixed by the cold dry mistral winds blowing down the Rhône Valley.

These joint French, US and UK MEDOC (Mediterranean Oceanography) experiments used the latest technologies. There were moorings and many CTD profiles. Neutrally buoyant floats were operated by NIO and by WHOI's Art Voorhis. He had devised a float with angled vanes that measured vertical velocities in the convection region. MEDOC'69 was notable as being the first cruise following a major refit of *Discovery* in which her internal layout of laboratories had been changed to fit her for the new age of electronics. It was also the first cruise with a shipboard computer and satellite navigation. This, coupled with the two-component electromagnetic log, greatly improved navigation and also gave an opportunity to map surface currents derived from ship drift.

In January 1969, Swallow carried out a preliminary survey of the likely convection region on HMS *Hydra*. It showed a doming of the density surfaces that pre-conditioned the water column so as to make deep convection possible. During the subsequent MEDOC'70 experiment we observed for the first time ever an ocean in which the temperature was uniform to within a few millidegrees down to 2,000 m.[7] But the weather that caused the mixing was severe and resulted in damage to *Discovery* when, even when hove to on station, a wave carried away the forward hydro platform and shattered portholes. The other ships, *Atlantis II* and *Jean Charcot* fared equally badly and several people suffered broken limbs.

The late 1960s was a time when much wider-ranging work on the interactions between the atmosphere and oceans was started. In 1967, Charnock and Tom Ellison attended a conference in Stockholm in preparation for the Global Atmospheric Research Programme (GARP). Charnock's proposal to the Royal Society and Royal Meteorological Society for an air-sea interaction experiment as a UK contribution to GARP was accepted. Mindful of the early work in the Isles of Scilly and Anegada, it was felt that at least three ships would be needed to allow the determination of horizontal gradients. In addition, accurate tracking of meteorological balloons by theodolite would be impracticable on ships, so new instrumentation would have to be developed. Starting in 1970, NIO and Southampton University, where Charnock had been appointed to the chair of physical oceanography, mounted a series of exercises named

Henry Stommel (L) and Allan Robinson (R) engaged in heated discussion at a MODE planning workshop in Bermuda in January 1973.

JASIN (Joint Air-Sea Interaction). These involved measurements of both the upper ocean and the overlying atmosphere from ships and buoys. In the 1970 trial involving the Dutch ocean weather ship *Cumulus*, RRS *Discovery*, and the survey ship HMS *Hecla*, a new radiosonde tracking system using radio navigation signals was evaluated around the nominal position of Ocean Weather Station Juliet (52° 30'N 20° 00'W).

Other air-sea interaction research in 1970 involved a new senior scientist, Alister Watson, who had joined NIO from ARL. The background to his transfer is documented in Peter Wright's book *Spycatcher.*[8]

The JASIN experiments continued beyond the timeframe of this book. Further trials in 1972 and 1977 led to the final experiment in 1978, now relocated to the Rockall Trough which was thought to offer a more homogeneous ocean region than the original Juliet site. Fourteen ships and three research aircraft were used during the two-month experiment which coincided with the launch of SeaSat, the first dedicated oceanographic satellite for which JASIN became the prime source of validation data.[9]

The Mid-Ocean Dynamics Experiment

The culmination of the study of ocean variability in the NIO-era was our involvement as the sole non-US participant in the Mid-Ocean Dynamics Experiment (MODE) in 1973. This effectively brought all the strands of ocean circulation research together. Its US proponents were John Swallow's longstanding friend and collaborator Henry Stommel and Allan Robinson who, as a student, had participated in the *Aries* experiment.

MODE revisited the area near Bermuda studied on the *Aries* and involved significant advances in float technologies. Long-range SOFAR neutrally buoyant floats had been developed by Tom Rossby and Doug Webb in the USA and were tracked throughout the NW Atlantic from shore-based listening stations the purpose of which was to locate the impact of test fired ballistic missiles.

A Royal Society Discussion meeting on Ocean Currents and their Dynamics convened by Dr Deacon in November 1970[10] was likely to have strengthened the case for UK participation in MODE. In the conclusions of his paper at the meeting, that described the much earlier *Aries* measurements, John Swallow wrote "*There may sometimes be advantages in combining Lagrangian (float) measurements with those from fixed arrays ...*". That is exactly what MODE did.

At NIO, Brian McCartney and Nick Millard had developed the Minimode transponding float system that, by using lower frequencies (around 6 kHz), could extend float tracking to ranges up to 100 km compared with the typical 5 km of ship-tracked 10 kHz floats. We had purchased new US vector-averaging current meters that were more immune to mooring

The main lab on Discovery *in the 1970s. The instrument in the foreground is the conductivity/temperature/depth (CTD) probe. Note the reel-to-reel tape recorder as a backup for recording CTD data.*

motion and we used a state-of-the-art CTD probe that Geoff Morrison had helped to build in Woods Hole working with Neil Brown.

During MODE, we mapped mesoscale ocean currents on scales of hundreds of kilometers, and monitored the westward propagation of these features representing the 'weather' of the open ocean.[11] On the MODE cruises, Dr Deacon operated the salinometer with typical dedication and excellent results.

MODE could be regarded as the point at which ocean circulation studies at NIO entered the mainstream of international marine physics research. In fact some wag quipped that the acronym stood for Modern Oceanography, Damned Expensive! The advent of the SOFAR float, and in 1978 the start of ocean observations from satellites, marked the points at which ocean observations started to be freed from the limitations of ships and enabled entire ocean basins to be studied synoptically.[12,13] We also had access to the precursor of e-mail (telemail) that made international communication and hence the planning of large international experiments much easier.

It was during MODE that Swallow gathered the scientific party together in the main lab on *Discovery* to read a telegram from the NIO Director, Charnock, giving details of how NIO would become IOS.

And what is the legacy of those formative NIO years? The international Argo programme uses 3,000 Swallow floats in a new temperature/

salinity profiling, satellite-tracked guise and provides year-round CTD profiles in undreamed of quantities (100,000 profiles a year) throughout the ice-free oceans. Argo is the central element of the *in situ* ocean observing system.[14] Moorings, based on the technologies developed in the 1960s by Mac Harris and Dennis Gaunt and deployed by Gaunt's winch are maintained year-round at 26°N to monitor the strength of the Atlantic meridional circulation.[15]

9

Internal waves and all that

Steve Thorpe

New Year 1962, and expecting to leave university in June, I was looking for a job. Harwell?–but I wasn't keen on radiation or on magneto-hydrodynamics. The Radar Research Laboratory?–but I knew nothing of radar or even wireless. The Road Research Lab?–but I couldn't even drive. And then (the kind Mr Coope in the University Appointments Board had just visited it), the National Institute of Oceanography was suggested. My love-hate relationship with the sea was challenged, or again forgotten (I'd 'enjoyed' seasickness in the Sea Scouts); it sounded as if it might offer an exciting job. I visited in early February and was 'shown round' by Dr Cartwright, Mr Crease and Mr Tucker (the Director, Dr Deacon, being unavoidably absent). After interview by a Royal Naval Scientific Personnel Board in London, I received a letter in April from the Secretary of the Admiralty, Bath: *"I am commanded by My Lords Commissioners of the Admiralty to offer you an appointment as a Temporary Scientific Officer in the RN Scientific Service ... at an initial salary of £848 per annum"*. This appeared satisfactory–a sufficient, if not magnanimous, salary, and 'Officer' sounded superior to my last rank, that of Corporal during National Service! A friendly letter followed from Mr Williams, Secretary of the NIO, mentioning digs near the Institute at 5 guineas a week, with bed and breakfast, evening meals and full board on Saturdays and Sundays, including laundry–with probably a reduction if I was away at weekends.

But at the end of May the Cambridge Tripos results were published: my performance in Part 3 of the Maths Tripos was enough to merit the offer of a place to do postgraduate research in the Cambridge Department of Applied Mathematics and Theoretical Physics (DAMTP). Compromise: Dr Deacon agreed that I should join the NIO but be seconded to DAMTP for a time sufficient to satisfy the university residence requirement for a PhD. A discussion followed between my proposed joint supervisors, Dr Michael Longuet-Higgins (NIO) and Dr Owen Phillips (DAMTP).

The former suggested 'The dynamics of the Gulf Stream' and the latter 'Breaking internal waves'. I simply listened. Michael eventually gave in and I'm still working on the internal waves!

Internal waves

Curious creatures, internal waves! They depend for their existence on the fact that the density of the ocean increases with depth–please think of a jar with a layer of light oil floating on water: waves can occur on the top of the oil layer *and* on the interface between the oil and the denser water. The latter are internal waves. The equivalent of the interface in the sea is the thermocline, where the water temperature decreases and water density increases rapidly as you go deeper. Internal waves can travel along the thermocline, much as the (surface) waves do on the sea surface. But where the water density increases very gradually with depth, internal waves can travel upwards or downwards, at an angle to the horizontal. (Those going downwards may transport energy from the sea surface where they are generated by winds.) This property is unlike that of 'surface waves' for they are constrained to travel horizontally along the water surface. Internal waves are also affected by the way in which the speed of the water changes with depth. Indeed if the speed changes sufficiently rapidly, instead of propagating, internal waves will grow, eventually breaking (as I shall explain later), and cause turbulence and mixing. The water in which they occur is then 'unstable', unable to retain its density and velocity structure. Internal waves may also break if, when strongly forced, they become sufficiently large, although the ways in which breaking occurs are different from the familiar plunging or spilling of surface waves as they approach shore. Internal waves are important because they mix or stir the ocean, and because they carry energy from where it enters the sea at its contact with the atmosphere–the sea surface. They were first perceived as a nuisance: oceanographers wanted to measure how the average density at a given depth varies horizontally, because they could then employ a theoretical relationship involving the Earth's rotation (the 'Geostrophic Relationship') to estimate the speed of the large-scale currents. And currents were much more difficult to measure than the density, and much more interesting since it was the circulation of the ocean that many oceanographers wished to discover; the beastly internal waves caused variations or 'noise' that made measurement of the average density hard to do.

Of all this I knew nothing in 1962, and indeed very little was known about internal waves at that time. The subject was mostly open to be discovered. No one knew how internal waves might break. My dissertation was completed in 1966 and consisted mainly of the development of a theoretical description of the shape of internal waves, and laboratory experiments that tested the theory. I had discovered rather little about their breaking. But what became apparent was that the waves produced in my laboratory tank were very different in their shape from those that had been measured in 1952 in the long, and relatively narrow, Loch Ness by Clifford Mortimer.[1] He had used a then newly-available chain of temperature sensors to measure the temperature variations in the loch, finding that the internal waves there appeared to have very steep fronts, whereas mine were more closely sinusoidal: as they pass, the waves on the thermocline of Loch Ness cause an abrupt increase in thickness of the upper layer of warmer water, so a sudden increase in the temperature of water measured at fixed depths as the warmer water from nearer the water surface is carried downwards. Clifford had only published measurements from one end of the loch, near Fort Augustus, where the abrupt jumps occur about 50-60 hrs apart. Surprisingly, measurements half way along the 34 km-long loch showed changes that were 25-30 hrs apart, half the 'period' at the end.

What was going on? Two waves in the middle for every one at the end? I guessed that the waves in Loch Ness were more like shock waves, travelling to and fro along the loch, passing the centre twice (and giving half the period) for every time they reached the end. Measurements were made with a thermistor chain borrowed from the Marine Science Labs in Menai Bridge, North Wales. The chain was moored near Urquart Castle, a quarter of the way down the loch, in autumn, 1970. Results proved that the increases in thermocline depth as the waves passed were often as great as 20 m but with no clear change to the appearance or the level of the water surface. The jumps arrived at successive intervals of about a quarter and three-quarters of the period found at Fort Augustus, just as expected as the 'shock' or jump travelled back and forth along the loch.[2] But it was also found that usually, rather than being an abrupt thickening of the warmer layer, the jumps actually consist of a series of waves, rather like those often seen following the tidal bore up the River Severn and referred to as an 'undular bore', meaning a change in water level with following waves. (With the expert help of Arnold Madgwick, the NIO Chief Photographer, and his able assistant, Norman Mansbridge, and with advice from F.W. Rowbotham of the Severn River Authority and Dr Howell Peregrine of the University of Bristol, a movie film was completed in about 1971 called 'The Severn Bore'. It follows the bore upriver from Awre until the weir at Gloucester, was successful in winning some movie awards, and later proved valuable in teaching undergraduates.) Internal waves of this kind, now commonly referred to as 'soliton internal wave packets' or 'internal solibores', are now known to occur frequently near the edge of the continental shelf where they are forced by tidal motions.

Because they smooth and roughen the sea surface they can be detected from orbiting satellites.

Experiments in a long tube sealed at each end, completely filled with two layers of water and denser brine and oscillated about a horizontal position, demonstrated the shock-like form of the internal undular bores in Loch Ness, and theory showed that they could be produced 'resonantly' by storms passing over the loch at intervals of 2-3 days, in a manner similar to the forcing of a swing pushed at intervals close to its natural oscillation period.[3] (The difference between the waves in Loch Ness and those I had studied earlier was one of wavelength compared to the depth; long waves can develop shocks or solitons. Analogous differences–but to 'surface' waves–are seen in a bath. In a deep bath, water 'sloshes' to-and-fro, but in a shallow bath, waves form 'jumps' that travel along, reflecting at the ends. Try it, but don't blame me if the water spills over!) As had been foreseen by Sir John Murray in seeking funds for his *Bathymetrical Survey of the Scottish Fresh-Water Lochs* at the end of the nineteenth century, the loch proved to be a natural laboratory in which to study many processes that occur in the ocean. Because its water is fresh, salinity does not contribute to its density which can therefore be relatively easily determined from measurements just of temperature, and numerous studies relating to mixing, made possible with novel instruments, including acoustics, were made there in the subsequent two decades, with the very helpful collaboration and technical advice of Mr Alan Hall and other members of the Institute.

Instability!

The long tube also proved to be a useful apparatus in which to make other experiments, particularly those relating to the instability of flows in the ocean. As well as having a density that increases with depth, the speed of seawater (i.e., the current) also changes in depth, sometimes so rapidly that small wavelike disturbances grow and become turbulent, causing mixing and changing the water's density and speed. The laboratory experiments hinge on being able to produce a slowly increasing and predictable flow. How this could be produced was discovered by accident, although it was later realised that Osborne Reynolds, famous for his description and demonstration of the transition from a smooth parallel laminar flow to a turbulent flow, had used the method in 1883! One morning in 1965, my supervisor in DAMPT, Dr Brooke Benjamin–Owen Phillips had by then departed to take a position at Johns Hopkins University in the USA–came to tell me that the previous evening he had taken home a transparent Perspex tube, and part-filled it with coloured brine overlain with a layer of water. Rocking the tube about an angle to the horizontal produced waves on the water-brine interface that appeared to break where the interface met the inclined sides of the containing tube: this might be a way to study breaking waves. I repeated the experiment in a 1.5 m-long tube with circular cross-section, although the tube was first placed in a horizontal position, allowing the dense brine to spread, making the interface horizontal. On slightly lifting one end of the tube to produce the tilted configuration that Brooke had used, I was astonished to see that after a few seconds a beautiful regular array of stationary eddies, now known as 'billows', formed in the tube, mixing the water and the brine together.

After repeating this several times it was realised that, when the end of the tube is lifted, the brine layer accelerates 'downhill' and the water runs 'uphill', replacing the void left by the brine, so creating a velocity along the tube that varied with depth. The rate of variation of velocity in depth, or 'shear', is greatest at the interface and can easily be calculated. When this shear becomes sufficiently large, the flow becomes unstable; very small internal wave disturbances caused in lifting the tube grow rapidly in amplitude. A theory was available to predict the wavelength of the disturbances that grow fastest and eventually form the billows in a steady flow, and this was developed to allow for the flow's acceleration.[4,5] The tilted tube itself is similar to the short tubes now commonplace as mobile ornaments filled with coloured liquids and made to oscillate, creating elegant, slowly evolving waves. A larger tube was constructed, about 6 m in length, with a rectangular cross-section and, using it, the

Top: *'Billows formed in a tilted tube filled with two uniform layers, one of fresh water and (below) of denser, coloured brine. The tube is 10 cm deep.*
Bottom: *Billows in the sky.*

theoretical predictions of the conditions for the onset of instability (known as 'Kelvin-Helmholtz instability' after the two great scientists who devised a relevant theory in the 1860s) were verified. Also studied were the structure of the billows and the properties of the mixing and the turbulent flow that resulted from the instability. The quantification of the instability provided numbers that proved useful in estimating the energy lost by internal waves when they produce sufficient shear to cause mixing. Kelvin-Helmholtz instability is now known to be widespread in the ocean. The observations of pattern and structure in the evolving billows lead to advances in understanding of the development of turbulence, a shady and still poorly-understood area of science, but one of broad application.

The research on billows was indeed recognised as having application to the prediction of atmospheric turbulence, particularly in clear air. (The billows of clear air turbulence or just 'CAT' can be detected using radar.) In September 1971, a talk, which included a demonstration, was given at the British Association Meeting in Swansea entitled '*A model for CAT and the Monster Internal Waves of Loch Ness*'. As hoped, the title attracted a large audience. The front row was however filled with the 'Press' and in spite of my denying that the work had any bearing on the Loch Ness Monster, the 9 pm TV News and next day's newspapers carried items saying that there was a new explanation for sightings of the Monster!

By 1970, I found myself a member of the Atmospheric Environment Committee (AEC) of the Aeronautical Research Council, a committee that provided a link between the research community and the military and civilian aircraft operators, and discussed in some detail, for example, how weather might have caused a particular aircraft accident, and how better warning of hazards might be made available to aircraft traffic controllers. On the committee were some of the leading intellectuals of the UK atmospheric science community, notably R.P. Pearce, P.A. Shephard and R.S. Scorer. Henry Charnock was also on the AEC. (It seemed a pity that no equivalent 'user' organisation concerned with marine applications of recent research seemed to exist.)

Flow near the abyssal seabed and the water of the Mediterranean

Much of what was done, and the freedom given to do it, was with the encouragement and support of the Director of the NIO, Dr George Deacon. Although rather as an 'aside' to my main research, one example may suffice to demonstrate the nature of his support. Dr Adrian Gill, the DAMTP co-organiser of the Cambridge-NIO Seminars, had derived a theory of a boundary layer on the seabed, the 'benthic boundary layer'. This layer is the zone in which the motion in the main body of the water adjusts to the condition of zero motion where it is in contact with the seabed. The direction of the flow in the layer is affected by the Earth's rotation: the speed decreases as the bottom is approached and the flow direction changes or 'spirals'. A theory of these spiralling changes to the flow was first developed by V.W. Ekman in 1905, but Adrian had extended it to include effects of turbulence. He predicted that the layer would be only a couple of metres in thickness. In the mid-1960s there was still much debate about whether such boundary layers with rotating current directions actually existed, and I therefore set out to see. Tony Laughton had developed an underwater camera and one was kindly made available. A tripod-like framework made from scaffold poles was designed with the camera mounted on the top, set to look vertically down a wire from which vanes, carefully weighted to be neutrally buoyant at the bottom of the Bay of Biscay, were attached.

When lowered to the seabed the photographs would, it was hoped, show a turning of the vanes with height above the bottom, the 'Ekman spiral'. The workshops had problems in completing the parts of the structure in time for the cruise. Dr Deacon heard of the delay and went down to the workshops and gave instructions that 'young Thorpe's' equipment must be given priority. I heard of this only much later, but it is characteristic of Dr Deacon's quiet and gentle, but firm, management and support for his scientists, and he was respected and loved the more for it. And there was also great professional support from the technical experts, the staff who designed and constructed equipment and who made the electronics and sensors to measure the ocean.

What of the Ekman spiral? The tripod framework (by then, because of its thin, if not elegant, legs, called 'Twiggy II'), heavily weighted to keep it in place on the bottom and equipped to release dye onto the seabed on arrival – the dye-plumes would reveal the flow very close to the bed – was lowered on the trawl warp from *Discovery* to the bottom of the Bay in November 1967, and recovered after about an hour or so. Having then no satellite navigation, it was impossible to tell whether or not the ship was drifting, hauling the rig along the seabed – and it clearly *had* been, for Twiggy II returned rather 'mangled', but with all parts, including camera, intact. Although the photographs taken whilst the rig was stationary and upright after reaching the bottom showed some variations in the direction of the flow with distance from the seabed, the evidence for a spiral was not convincing. Transient variability, not least the tides, was not measured or 'averaged out' by the short period of the observations. Twiggy II was later deployed for periods of several days in March 1971, on a mooring in the Gulf of Cadiz. This revealed the variations and strength of currents in the relatively warm and salty water flowing from the Mediterranean and, quite unexpectedly, discovered a cloud of sediment flocs (small clusters of sediment particles, rather like snowflakes, about 2 mm in size),

'Twiggy II', a framework designed to measure the variation of current direction close to the seabed.

extending from below the outflow to the seabed, a cloud some 200 m thick. Measurements of turbulence near the seabed in the Gulf were also made in 1970 and 1971 using an electromagnetic current meter.[6,7]

When first visiting the NIO I saw a bottle labelled 'Mediterranean Water'. A joke? Had someone brought back a bottle after a holiday in the warmth of the Côte d'Azur? No! It was solemnly explained that the water has particular chemical properties that identify it, and that a layer of water of Mediterranean origin is found at a depth of about 1000 m in much of Northeast Atlantic, providing clues to ocean circulation. The purpose of the studies in the Gulf was to understand just how the salty Mediterranean water gets to this depth, and what dilution and mixing occurs as it spreads into the Atlantic after escaping through Straits of Gibraltar.

It was a golden time of exciting discoveries and of wonderful opportunities, with far greater freedom for research and exploration than there is now.

10

Seawater – its chemical and physical properties

Fred Culkin

The chemistry of seawater is fundamental to nearly all other aspects of oceanography. In particular, the total salt content, or 'salinity', together with temperature and pressure, govern the density of seawater and a knowledge of the distribution of this is fundamental to our understanding of the circulation of the oceans. Dr Deacon was well aware of this as a result of his work in the Southern Ocean and so one of the early appointments to NIO (in 1950) was a chemist, Roland Cox. As has been said elsewhere, the organisation of NIO was very informal and, as other staff joined Cox, NIO acquired a *de facto* Chemistry Group.

The problem with determining salinity is that high accuracy is required and this presents many problems. Before discussing these and their solutions it is necessary to explain briefly the basic chemistry of seawater.

Defining and measuring salinity

Salinity was originally (ca. 1900) taken to mean the total quantity of dissolved salts (with certain chemical reservations) in 1 kg of seawater (defined in parts per thousand by weight ‰). It may still be considered as such, though a change in the definition to one related to electrical conductivity has altered its numerical value slightly and led to it being defined as just a number with no units attached. Though minor constituents such as phosphate, silicate, nitrate and dissolved oxygen occur in concentrations of micrograms/kilogram, they are important because they control biological productivity. Thus they are referred to as nutrients. The major contributors to salinity are sodium, potassium, calcium, magnesium, strontium, chloride, sulphate, bromide, bicarbonate, boron and fluoride, and these occur in almost constant proportions with only small, but detectable, regional and depth variations.

The most accurate physical work and especially that in the deep ocean requires a precision of better than 0.002 in salinity. Since the salinity is typically in the region of 35, this implies an accuracy of about 1 part in 20,000 of the salt content. However, because it is impossible to determine salinity with the required precision by measuring dissolved solids, indirect methods have been used over the past century. Approaches have included the determination of electrical conductivity, optical refractive index, direct measurement of density and the velocity of sound, but until the 1960s the most useful method was chemical.

A fortunate consequence of the near constant proportions of seawater's major constituents (in the open ocean) is that the total salts can be estimated from the concentration of any one. Chloride can be determined very accurately by drop-wise titration with silver nitrate solution. The concentration of the total halides (chloride, bromide and iodide) is called chlorinity. At the end of the 19th century, Danish workers determined the total salt content of a number of seawater samples by evaporating the water and weighing the residue to get the salinity (S) and they then determined their chlorinities (Cl) and found a linear relationship between the two.

In 1900, Martin Knudsen, working in Copenhagen, prepared ampoules of seawater on which he determined the chlorinity very accurately. He distributed some of these ampoules to other laboratories for use as a standard in chlorinity titrations, thus founding what later became known as the Standard Seawater Service. Titrating the chlorinity and converting this to salinity by use of simple equations which had been determined for a few seawaters was the standard method for over half a century.

However, chlorinity titration was time-consuming and difficult to carry out at sea even by a skilled chemist; it was capable of an accuracy (precision) of only about 0.02 in salinity. 'Only' is perhaps a modest statement as titrations of any sort to this precision at sea or in the lab were among the most precise carried out routinely anywhere. Consequently, in the 1950s, efforts were made to replace titrations by measurements of electrical conductivity, a property strongly dependent on salinity.

Roland Cox on the poop deck of RRS Discovery II.

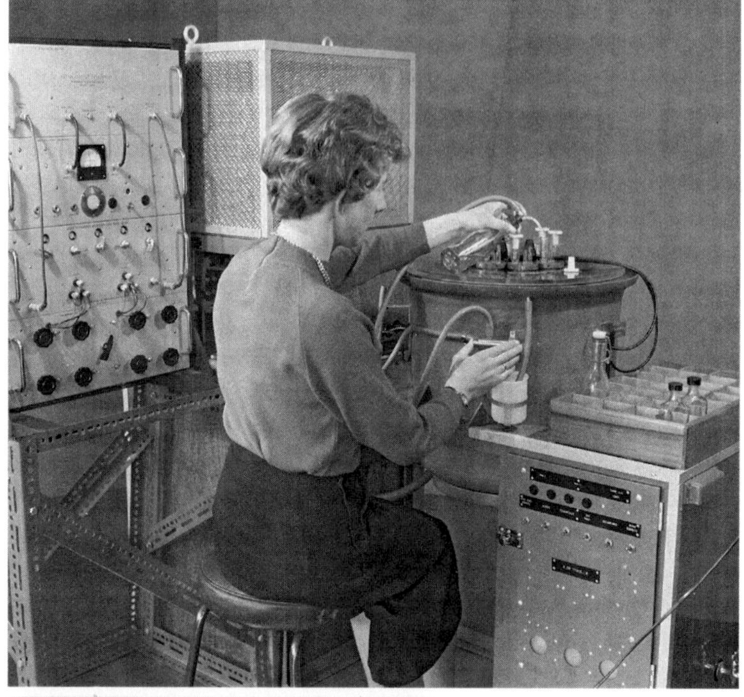

Above: *The NIO Thermostat Salinometer in use. The operator is Roland Cox's wife, Audrey.*

Left: *Detail of the bath containing the measurement cells.*

Thermostat salinometers were developed that held the samples in a temperature-controlled bath and used a high-precision comparator bridge to compare the electrical conductivity of seawater with that of a standard. Unfortunately, although these salinometers were capable of high precision at sea and on land, the basic data for converting conductivity to salinity were inadequate. It was the work of NIO and a few other laboratories that led to an order-of-magnitude improvement in precision and opened the way for the study of structural variations on a smaller scale than the ocean basin.

Roland Cox was one of the first scientists to attempt to improve our knowledge of seawater. His early innovative work included the measurement of its specific heat (with Norman Smith),[1] a method of determining salinity using a compressible float and the design of the NIO Thermostat Salinometer (with Tom Tucker and Ray Peacock). This was widely used in the 1950s and 1960s.

Cox also became aware of the unsatisfactory knowledge of the physical and chemical properties of seawater and, in 1959, he embarked on an ambitious investigation of the relationships between chlorinity, electrical conductivity, specific gravity, refractive index and temperature.[2,3,4] This project and its successors were to be of fundamental value to the understanding of small-scale processes in the ocean. With the help of many laboratories about 400 samples of seawater were collected at various depths in all the major oceans and shipped to Wormley where they were filtered and sealed in glass ampoules.

I joined NIO in 1960 and after a cruise on *Discovery II* on which I gained my first seagoing experience of the NIO Thermostat Salinometer, I joined Cox's programme. I measured the chlorinity of most of the 400 samples, using a high-precision titration method[5] adapted from that used in Copenhagen for calibrating Standard Seawater, and also the associated conductivities at 15°C using the NIO thermostat salinometer. Cox himself studied the effect of temperature on conductivity nearer to 0°C, frequently working overnight with the salinometer in his garage at home to obtain lower ambient temperatures. To complete the dataset, Mike McCartney measured specific gravities relative to an isotopically-defined pure water and Stuart Rusby[6] measured optical refractive index which had been suggested as an alternative to conductivity for estimating salinity.

While making these measurements we established a close relationship with Frede Hermann who was Director of the IAPSO (International Association of Physical Sciences of the Ocean), Standard Seawater Service based in Copenhagen. His responsibility was to prepare, calibrate and distribute Standard Seawater as the internationally recommended standard for salinity determinations. He invited me to carry out check determinations on each batch. This gave me experience of the high-precision chlorinity titration and ultimately led to my taking responsibility for the operation of the Standard Seawater Service and its transfer to Wormley in 1974.

One of the major findings of the NIO investigation was that there was significantly more scatter in the relationship between chlorinity and specific gravity than between conductivity and specific gravity. This was not surprising since chlorinity measures only one constituent whereas conductivity involves all dissolved constituents. The conclusion was that for the precise estimation of density, which is one of the objectives in measuring salinity, it was better to use conductivity than chlorinity.

The method of determining the specific gravity adopted at NIO was the classical one of comparing the density of seawater with that of pure water at 4°C (temperature of maximum density of pure water) and it required a pure water of known isotopic composition, since variations in the ratios of hydrogen and oxygen isotopes affect the density. No data on these effects were available to the required precision. This problem was partly overcome at NIO by preparing pure water from deep Mediterranean seawater by controlled distillation without changing the isotopic composition (isotopic analyses were carried out elsewhere).[7] The NIO specific gravity data proved to be consistent with some previous determinations, and although of higher accuracy, it was decided that no new density tables should be published until further measurements at low temperatures and higher pressures had been made by a new method (in 1969) based on the measurement of the characteristic vibrational frequency of a glass capillary filled with liquid.

A proposal was made to remove the anomaly that the standard used to calibrate salinometers in conductivity was IAPSO Standard Seawater which was still certified only in chlorinity. A possible alternative was to calibrate Standard Seawater in absolute electrical conductivity. In principle this meant measuring the resistance of seawater in a conductivity cell of known dimensions. Cox, in collaboration with Charles Clayson and Norman Smith, designed a rectangular quartz cell of uniform cross section. Several versions of this instrument were thoroughly tested and it proved capable of the required precision but the method was too time consuming for routine calibration of Standard Seawater.[8] We did, however, find it ideal in later work with other laboratories leading to the formulation of the Practical Salinity Scale 1978 and the International Equation of State of Seawater 1980.[9]

International standards

In 1961 the Intergovernmental Oceanographic Commission (IOC) of UNESCO was established in Paris and the need for standardisation, particularly in methodology, was widely recognised. Cox reported the initial NIO findings to the Hydrographic Committee of ICES and following pressure from ICES, and at the instigation of Cox, a Joint Panel on the Equation of State of Seawater was formed in 1962, with representatives from ICES, SCOR and UNESCO. ('Equation of State' refers to the relationship between the physical properties of seawater and the effects on them of temperature and pressure). But in 1963 the panel's scope was widened to include a request to carry out the preparatory work for publishing new oceanographic tables. The panel became the UNESCO Joint Panel on Oceanographic Tables and Standards (JPOTS).

The JPOTS made a number of recommendations based on the NIO data:

(1) the chlorinity data should be converted to salinity by the equation $S = 1.80655\,Cl$ (which replaced the old Knudsen equation $S = 1.805Cl + 0.030$);

(2) the new salinity data should be combined with the NIO conductivity data to form Volume 1 of the International Oceanographic Tables tabulating conductivity as a function of salinity and temperature; and

(3) chlorinity should be regarded henceforth as a property of seawater, independent of salinity, with no one-to-one relationship between the two.

The computations were made by Jim Crease and Pam Edwards and the tables were published jointly by NIO and UNESCO in 1966.[10] A second volume based on the measurements of refractive index anomaly by Stuart Rusby was published in 1967.[11]

Cox played a major part in setting up JPOTS, and in its early activities, and his death in 1967 at the age of 44 was a severe blow. However, the results from the ambitious programme that he initiated, and the new tables which followed and were published after his death, did much to introduce uniformity in salinity determinations. Subsequently, I replaced Cox on JPOTS and Crease acted as an invited expert. The Panel's work led to a new Equation of State of Seawater in 1980.

Although Volume 1 of the International Oceanographic Tables proved invaluable for laboratory measurements of salinity, the introduction in the 1960s of probes that measured salinity/conductivity, temperature and depth *in situ* produced new requirements. The Tables covered salinities from 2.5 to 41.5 and temperatures from 13.8 to 28.7°C, which was adequate for laboratory work. But for conversion of deep sea *in-situ* measurements, another dataset by Brown and Allentoft from 0 to 60 in salinity and 0 to 30°C had to be used.[12] The two datasets were not completely compatible because the International Oceanographic tables were based on natural seawaters while the other set were based on seawaters prepared by weight dilution or evaporation to give a range of salinities and hence had a slightly different chemical composition. JPOTS felt that another set of measurements should be made on a range of salinities prepared by dilution with pure water or evaporation of one seawater so that all the samples would have the same chemical composition. For the starting seawater a single batch of Standard Seawater was chosen and distributed to several participating laboratories for measurements of electrical conductivity.

Calibration of Standard Seawater to the required precision in absolute electrical conductivity, instead of chlorinity, was not practicable. This led NIO to be involved in the solution of another problem. Standard Seawater certified in chlorinity was satisfactory as a standard for use in the titration method because reaction with the glass was unlikely to produce a change in calibration as glass contains no halides. However, storage in glass was less satisfactory for a conductivity standard because dissolution of the container would certainly change the conductivity. There was a need for a well-defined electrolyte solution that could be prepared when required and so would be independent of any storage method. To maintain continuity with the previous salinity scale, the concentration of this solution should be such that its electrical conductivity was the same as that of a seawater of salinity exactly 35 at the same temperature and pressure.

The most suitable electrolyte was potassium chloride (KCl) which was the accepted conductivity standard in electrochemistry and was readily available in a state of high purity. JPOTS asked NIO and two other laboratories to determine the required concentration and we prepared sealed ampoules of filtered seawater for the investigation. Each laboratory developed its own method and Norman Smith and I used the apparatus designed for measuring the absolute conductivity of Standard Seawater. This gave high-quality data on the absolute conductivity of a range of KCl solutions and of seawater of salinity 35. The NIO figures were in excellent agreement with those of the other two laboratories (in Canada and France). This work and that of several other laboratories on variation of conductivity with temperature and pressure, led to a Practical Salinity Scale in 1978 and a new Equation of State of Seawater in 1980.[9] Thus the NIO aims of the 1950s and 1960s were finally achieved.

Seawater chemistry

The collection of 400 seawater samples, representative of most of the major oceans and seas allowed me to carry out some interesting chemistry. I looked at the relative proportions of major elements, a topic which had not been studied on any large scale since the *Challenger* expedition. The existing analytical methods for the major seawater elements, with the exception of chloride, were neither easy to use nor accurate when applied to seawater because of interference of some elements in the determination of others, so new methods had to be developed. The concentrations of elements per se are not very interesting, as they vary with the salinity, so it is usual to calculate the ratio to chlorinity (which is not affected by biological activity). The results showed a tendency for calcium to increase, relative to chlorinity, with depth, probably because of dissolution of calcareous debris at the sea floor. Variations in the other elements were very small, however, and mostly within the limits of the methods.[13]

The new analytical methods were useful in a study of the Red Sea brines sampled when RRS *Discovery* was returning from the Indian Ocean Expedition. At NIO we studied the major constituents and the minor constituents were analysed at Liverpool University. Surprisingly the brine proved to be very different in many respects from seawater that might have been in contact with brine/salt deposits or subjected

Jim Crease drawing samples during the IIOE from the new (bright yellow) NIO water bottles. The metal tubes on the front hold thermometers.

to evaporation. We concluded that the high temperature and unusual composition pointed to the water having a source originating from water in old sediments from which it was removed by geothermal heating emerging as a hot spring. Since then, highly saline, high-temperature waters of different origin have been found in other oceanic areas but this was probably one of the first examples to be subjected to such detailed chemical analysis.[14]

Uncontaminated sampling of seawater is fundamental to marine chemistry and before the IIOE we needed to replace the old metal Kullenberg water sampler with something larger and which would not introduce (metallic) contamination. Since no such sampler was available commercially the Engineering group designed the NIO Water Sampler made in polypropylene. After extensive testing it was used for the next 20 years.

Salinity was not the sole focus of the Chemistry group. Chemical analysis was needed to study nutrient cycles and the mixing processes in which nutrients were involved. The standard analytical methods for nutrients involved treating the sample with chemicals to convert the nutrient to a coloured compound, the intensity of which was related to the concentration. For example, phosphate is treated with ammonium molybdate and then a reducing agent to produce a blue phosphomolybdate complex. Commercial spectrophotometers that measured quantitatively the intensity of the colour worked well on land, but most of them proved electronically unstable at sea. This had to be improved before the IIOE and the Chemistry Group put considerable effort into modifying and testing the most promising commercial instruments. A short cruise on *Discovery II* in the English Channel compared five instruments and showed that the main cause of instability was flexing of the prism control arm which needed strengthening to make a more satisfactory seagoing instrument.

While much of NIO's chemistry supported physical and biological programmes there were independent chemical programmes. The presence of biologists who could identify and supply samples enabled us to work on the composition of marine organisms and in 1967 we started a study of the fatty acid composition of the lipids in a range of marine animals.

Lipids are complex mixtures of long-chain (between 12 and 22 carbon atoms) fatty acids and their compounds with sterols, glycerol and phosphoric acid. When Gas Liquid Chromatography (GLC) became available in the 1950s and 1960s it became possible to study them in small samples. Volatile derivatives from the samples were passed through a heated chromatographic column, which separates the components in order of their molecular weights so that their relative abundance can be recorded. To run this work in 1968 we recruited Bob Morris and since much published data on marine lipids referred to fish and mammals we started to look at the smaller zooplankton including euphausiids, decapods, small fish, and small cephalopods. In some cases, it was possible to correlate fatty acid composition with diet. Two species of filter-feeding tunicates, for example, were found to contain unusually high levels of myristic acid (C-14) that is characteristic of their phytoplankton diet. Field studies of the effects of environmental and seasonal factors on lipid composition were continued under IOS.

One piece of marine chemical work did not find its way into the literature. I received a package from a colleague at the Sea Mammal Research Unit in Cambridge, containing about 200 g of a pale yellow crystalline material, which had been caught in a Scottish fisherman's net. The optimistic finder thought it might be ambergris, the fatty secretion from the intestines of whales, which is used in perfumery. It was quite valuable. Undaunted, but with increasing trepidation I followed the procedures for the identification of unknown substances that had been part of my chemistry degree course. Recrystallisation produced lovely crystals, still yellow, suggesting a nitro compound. Fusion with sodium metal gave a blood red colour indicative of a polynitro compound, but which one? The melting point, 80.1°C, and a couple of derivatives led to the inevitable conclusion that it was that well-known explosive 2:4:6: trinitrotoluene (TNT). I hastily rendered the sample harmless and sent a letter to the finder advising him to take the rest of his find to his nearest police station for safe disposal!

The Visible Surface of the Ocean

Waves are the most visible manifestation of the movement of the waters of the seas. Group W laid the basis for the theoretical understanding and the measurement of waves which continued at NIO. Michael Longuet-Higgins, a mathematician with strong connections both in Cambridge and the USA, gives the theoretical basis to how waves are generated by wind and how they can be predicted. David Cartwright, later an Assistant Director of IOS in charge of the Bidston laboratory, writes of the development of the means of measuring waves at sea, of storm surges (such as the devastating one in the North Sea in 1953), and of oceanic tides. Finally 'Tom' Tucker, later Assistant Director at IOS Taunton, shows how the theory and measurement of waves have important practical applications in ocean engineering, especially in the design of offshore oil and gas installations.

11

Wave research at Wormley

Michael Longuet-Higgins

Throughout the 1950s and 1960s, NIO maintained the high reputation for wave research which had been gained from the work of Group W. Four main aspects were combined; basic theory, laboratory experiments, field observations (often involving new instruments and techniques) and practical applications. These were all interlinked. NIO maintained close connections with other leading research establishments and universities. The support given by NIO's technical experts allowed theories to be tested.

Wind and waves

In 1954, when I rejoined NIO at Wormley, the question of how the familiar waves we see on the ocean surface are actually generated by the wind was largely an unsolved problem. Rough rules that served the needs of sailors were available to be sure,[1] and much light had been thrown on the propagation of swell outside the storm area by Harald Sverdrup and Walter Munk in California[2] and by Group W. In 1956, the situation was summed up by Fritz Ursell (then in the Department of Applied Mathematics in Cambridge) in a famous review article:[3] "*Wind blowing over a water surface generates waves in the water by physical processes which cannot be regarded as known*".

This then was the problem that I set out to attack. The first task was to obtain reliable data from the storm area itself. Some of the tools were ready to hand. In Group W, Norman Barber, in collaboration with Norman Smith and Frank Pierce, had already designed a 'pitch-and-roll' buoy that could simultaneously measure and record the vertical motion of the sea surface, using an accelerometer, and the tilt of the surface in two directions at right-angles (say N-S and E-W) by means of two gyroscopes. From such data, Barber in 1946[4] had shown theoretically that one could obtain some indication of the direction of travel of the waves, in any given band of frequencies. While at Cambridge in 1955, I had further developed

The pitch-and-roll buoy being deployed from the Discovery II.

Barber's analysis to show how to obtain both the mean direction of travel and a measure of the angular spread of directions.

The pitch-and-roll buoy had been brought from Teddington to Wormley during the move in 1953. With the help of Norman Smith, it was prepared for sea, and a new feature was added; a microbarograph for measuring the pressure in the airflow over the buoy. The buoy was carefully calibrated for its response to waves of different frequencies in the new long wave channel at the National Physical Laboratory in Feltham, with the permission and help of Neil Hogben.

The idea was to find whether the pressure was systematically greater on the windward side of a wave crest than on the lee side. This would be a critical measurement to decide between different mechanisms of wave generation. In July 1955, Norman Smith and I took the pitch-and-roll buoy to sea in *Discovery II.* For recording storm waves, the higher the wind, the better. The upper limit was set by our ability to launch and recover

The NIO 'clover-leaf' buoy.

the buoy from the heaving deck of the ship. One good storm was missed and then the chief scientist, Henry Herdman, developed peritonitis and had to be landed at Cobh, where he was taken to hospital. Fortunately we encountered another Atlantic storm further south. Norman Smith seemed almost immune to sea sickness and was able to work on maintaining the buoy in the roughest weather in which we could operate (Force 5).

After a long period of analysis, in which David Cartwright joined us using new, digital methods of Fourier analysis which replaced the old rotating-drum wave analyser, we presented our results in 1961 at a conference on ocean waves convened by the Office of Naval Research (ONR) in the USA.[5] It was held at the little town of Easton in Maryland. We had measured the energy spectrum of storm waves, and found it to be nearly in agreement with a power-law proposed in 1958 by Owen Phillips,[6] though it was even closer, as it later turned out, to the revised theoretical spectrum that he published in 1985.[7] We had determined the angular spread of the energy for given wave periods and shown that it diminished as the wave period grew shorter. Only the measurements of air pressure were problematical, but our pioneering efforts were later improved on by other workers. Our results were at least consistent with the theories of Miles[8] and Phillips.[9]

A spontaneous comment, recorded in the proceedings of the Easton conference, came from C.S. 'Chip' Cox, of the Scripps Institution of Oceanography: "*I would like to say that I consider this work a monumental undertaking, which is giving people who are working in the field of waves a profound sense of admiration for the work at NIO*".

The pitch-and-roll buoy was so successful that we decided on a more sophisticated development, the 'clover-leaf' buoy. The buoy was, in essence, three pitch-and-roll buoys linked by a triangular frame. It let us measure changes not only of the two components of slope, but also of the three components of curvature of the sea surface. The aim was to find the directional distribution of the wave energy with greater accuracy. The buoy was built and instrumented largely by Norman Smith. It was later deployed in a joint US-British study of the behaviour of ocean waves as they crossed the Gulf Stream, and also on an oil tanker on its way to the Middle East. Later it was used extensively by Japanese researchers who measured the angular spread of waves under different sea conditions.

Wave prediction and wave interaction

During the 1950s and 1960s, a great advance in the science of ocean wave prediction was made by the discovery of nonlinear wave interactions. In deep water, very low surface waves – those with extremely small slopes – do not interact; they simply propagate independently and can cross one another without alteration. In scientific terms, the linear theory works well for long, low swells. However, in 1960, it was shown theoretically by Owen Phillips,[10] who had by then moved from Cambridge to Johns Hopkins University in Baltimore, USA, that four separate surface waves, when their wavelengths and directions have certain values, can 'resonate' resulting in an exchange of energy between them. A calculation of this energy exchange was made in three classical papers by Klaus Hasselman in1962-63,[11,12,13] and had been presented at the Easton conference.

The calculation was impressive but extremely complicated. I therefore sought some kind of experimental test. I had access to one of the long wave channels at the Hydraulics Research Station at Wallingford, thanks to the kind permission of my colleague Mr R.C.H. Russell. A wave of length 1 m was generated at one end of the tank, followed by another of 9/4 times the length which, since it travelled faster, overtook the other. According to the theory this should have produced a wave with a length nine times the first travelling in the opposite direction. But apparently it did not: no 'reflected' wave was detectable. Puzzled, I worked out the coefficient of the interaction and found that it was exactly zero for any two initial waves propagated in the same direction; thus the theory was not disproved. However in the next simplest situation, for two initial waves at right-angles, the interaction was not zero, and should be measurable. So I approached the Admiralty Experimental Works at Haslar where there was a much smaller rectangular tank suitable for our purpose.

We arranged that along one side of the tank a hinged wavemaker generated waves of frequency f_1 in one direction. Along an adjacent side, a similar wavemaker generated waves of frequency f_2 travelling in a direction at right-angles to the first. A vertical wave probe then sampled the fluctuating elevation of the water surface at various points across the tank. The interaction was predicted to generate an oblique wave with frequency $(2f_1-f_2)$.

I had been making model cars for my young children. So, to extract from the wave record at each point the harmonic component of frequency $(2f_1-f_2)$ we used a differential gear like that between the driving wheels of a car: if the cage of the differential is rotated at a frequency f_1 while one wheel is rotated at a frequency f_2 then the other wheel rotates at a frequency $(2f_1-f_2)$. Such a device, was conceived during a car journey between Wormley and Haslar, and was subsequently built out of Meccano parts.

We found that the third wave grew in amplitude across the wave tank, just as the theory predicted. Our results were published in a paper in 1966 with Norman Smith as a co-author.[14] Our result was later confirmed in larger-scale experiments by McGoldrick and co-workers at Johns Hopkins University.[15]

Radiation stresses and some practical applications

An important theoretical development, which gave rise to many practical applications, was the discovery of the importance of the flux of momentum in water waves. The flux of energy in surface waves had long been familiar from textbooks such as Lamb's *Hydrodynamics* (1932). Fluxes of momentum, which are behind many wave phenomena, were much less appreciated.

My interest in this began when I was on a visit to Vancouver in 1961. Bob Stewart, a Canadian who had studied turbulence under George Batchelor at Cambridge, was supervising an old Danish sea captain named Drent, who was studying for an MSc during his retirement on Vancouver Island. (He was 80.) Drent, following a previous calculation in 1947 by a British engineer called Unna, had done a theoretical calculation of the steepening of short waves riding on longer gravity waves: they were both shorter and steeper at the long-wave crests than at their troughs. The shortening at the crests was not in doubt: it was due to a purely kinematic effect whereby the coming-together of the water particles to form the crests of the long waves causes the water surface to contract horizontally. However, the effect on the amplitude of the short waves was more subtle.

By calculation, I found that Unna's and Drent's answer was incorrect. Bob Stewart and I traced the physical reason to Unna's neglect of a basic interaction between the orbital velocities in the long waves and the flux of momentum in the short waves.[16] This flux of momentum put extra energy into the short waves and made them steeper near the long-wave crests, and less steep in the troughs. (There was an additional cause: the long-waves have an extra vertical deceleration at their crests, and this changes effective gravity for the short waves.) We showed that the short-wave momentum flux could be described in terms of a radiation stress. This is analogous to the stress exerted by light or other electromagnetic radiation when it is reflected from a plane surface.

It is the momentum flux from a jet engine that produces its thrust: the momentum flux from a hose pipe aimed at a flat surface produces the pressure on it. The radiation stress in an oscillating wave is simply the extra average momentum flux due to the wave motion.

Stewart and I showed that the radiation stress produces many other effects. For example when swell approaches a shoreline, then the rigidity of the bottom causes the long waves to shorten and steepen before they break. But the radiation stress causes the mean level of the surface to fall locally, producing a so-called 'wave set-down'. On the other hand, once the waves have broken, that is to say once they are inside the surf zone, the momentum flux toward the shore causes a much larger 'wave set-up'. I later demonstrated this very simply by laying a transparent tube on the beach, at right-angles to the shoreline, and observing the vigorous flow seawards inside the tube.[17]

A similar explanation was found for alongshore currents. These are the currents flowing parallel to a straight shoreline which occur when waves are incident at an oblique angle. They had been shown to be driven by the shorewards flux of alongshore momentum.[18] In a 1970 paper, I developed

Photograph of MV Bencruachan *showing her back broken by a freak wave in the Agulhas current in 1973. (By courtesy of Ben Line archives.)*

the theory and found very good agreement between the theory and careful observations in a laboratory wave tank.

Naturally, alongshore currents are closely associated with the onshore and alongshore transport of sand and sediment by ocean waves. This is of great practical importance for coastal engineers. For example, offshore breakwaters are sometimes built parallel to the shore with the intention of reducing the waves on their leeward side. But in reducing the wave height, the radiation stresses are also reduced, which interferes with the natural alongshore flow of sand and hence produces a pile-up of sand between the breakwater and the shoreline.

Another application of radiation stresses is to the occurrence of 'rogue waves', which batter and sink several ships per year in various parts of the ocean. Some types of rogue waves occur when wind-generated waves run into an opposing stream, such as the Agulhas current off SE Africa. This area is notorious as a graveyard for wave-battered vessels. The same effect is found for wind-waves in opposing tidal streams: every sailor knows that waves running against a tide produce dangerous sea conditions.

The concept of radiation stress can be applied to many different types of wave. In an expository article with Stewart in 1964 we explained the stresses in gravity waves and extended the analysis to short 'ripples'. This is the term commonly used for surface waves having wavelengths of a few centimeters or less, which are partly or wholly governed by surface tension. In 1963[19] I showed how the radiation stress in ripples can account for the amplitude of 'parasitic capillaries', the name given to those short capillary waves often visible on the forward face of steep gravity waves. These are important dynamically since they extract energy from the longer waves just as they are about to break. The parasitic capillaries can even prevent the longer waves from breaking.

In internal gravity waves, namely those found on the interface between layers in stratified fluids, the radiation stresses are even more important, since the particle velocities in these waves are often large compared to the phase velocity. Hence they tend to be more nonlinear.

Rossby waves

Large-scale motions in both the atmosphere and the ocean are strongly influenced by the rotation of the Earth. There is an important class of wave-like motions, whose frequency is proportional to this rate of rotation. If the Earth's rotation were to vanish, the frequency of these motions would tend to zero and the waves would be reduced to steady currents. Such waves are commonly called 'Rossby waves', though some authors prefer to use Rossby's original name for them, which was 'planetary waves'.

Carl Gustav Rossby (not be confused with his son Tom who is a Professor of Oceanography at the University of Rhode Island, USA) was a Swedish meteorologist who during WWII was at the University of Chicago. His discovery of planetary waves in the atmosphere was based on an insight due to Hans Ertel, in Germany, that the motions of fluid in a thin layer (such as the atmosphere) on a rotating globe are governed by the 'spin', or vorticity, of the fluid, divided by the fluid depth. This quantity is usually called the 'potential vorticity' and it tends to remain constant. Thus, if the depth of a column of water were to increase, so that a vertical column of water became longer and thinner, then the rate of rotation would have to increase, like an ice skater whose arms are drawn in. Rossby pointed out that in an atmosphere of constant depth, if a column of air in mid-latitudes is displaced southwards, then its spin relative to the rotating earth, must increase; if displaced northwards, its spin must decrease. This leads to a class of waves whose phase velocity is always towards the East.

Near the equator, the vertical component of Earth's rotation vanishes, and gravity becomes relatively important. Hence, in equatorial regions, as Maurice Rattray at the University of Washington had shown in 1964,[20] there are three possible waves; one (gravity-type) travelling westward, the second (gravity-type) travelling eastward and one eastward-travelling Rossby wave. In the ocean, if we take account of the thermal stratification, then we find internal waves of the same three types.[21] It is the eastward-travelling internal waves that accompany an El Niño episode; the phenomenon in the Pacific that drastically affects global weather and climate.

In the early 1960s, I began to apply some of what I had learned about surface waves in the ocean to both gravity and planetary waves in the ocean and atmosphere. In a first study in 1964[22] I showed that in mid-latitudes, although the phase-velocity of planetary waves was always to the east, the group-velocity, i.e. the velocity with which the energy propagates, can be either eastwards or westwards. The energy in longer waves, such as those generated by typical weather systems, always travels toward the west. This was the basis for James Lighthill's 1969 theory of 'westward intensification' of ocean currents:[23] the waves reflected from a western boundary have a smaller (eastwards) group velocity, hence their energy density is higher, producing jets like the Gulf Stream in the N Atlantic or the Kuroshio in the western Pacific.

In 1964, I studied theoretically the propagation of planetary waves over a complete sphere, and showed that the energetic zone for such waves must always lie between two latitude circles, north and south.[24] At each of these critical latitudes was a 'wave caustic' at which the planetary wave amplitude should be relatively high. The neighbourhood of these critical latitudes has since then been a subject of special field investigations.

In a third paper,[25] I studied the effect of ocean boundaries on planetary waves. I showed that the normal modes of oscillation should take the form of eastward-travelling carrier waves, modulated by an envelope which had zero amplitude on the boundary. Later, I solved a similar problem but for planetary waves over a complete hemisphere having longitudinal boundaries through the poles.[26]

The logical extension of these studies was to allow for the effects of both rotation of the globe and for gravity, on the oscillations of a thin layer of fluid of uniform depth. The problem for a complete sphere was solved in its entirety in a classical paper.[27] The mathematical model, as applied to the atmosphere, is a considerable idealisation. Nevertheless it has been a reference point for many later calculations.

In 1968, a visiting Canadian post-doc, Stephen Pond and I developed a similar type of model, but for a hemispherical ocean basin with boundaries passing through the north and south poles. This showed the presence of equatorial-type waves travelling eastwards along the equator as described above. On meeting the eastern boundary of the ocean their energy was converted into Kelvin waves travelling polewards along the eastern boundary. A Kelvin wave is a gravity wave which becomes trapped by the Earth's rotation so that the energy clings to a boundary. As it travels along the boundary it always has the boundary to its right in the Northern Hemisphere, to the left in the Southern Hemisphere. (Such waves are found in the North Sea, for example, where a storm surge generated north of Scotland tends to be propagated southwards down the east coast of England until it reaches the Thames estuary, then it turns and travels anti-clockwise round the North Sea.)

These models were merely the beginning of a programme of theoretical research designed to isolate and understand, one by one, the main factors controlling time-varying ocean currents. The next factor to be considered was the strong influence of topography; the changes in ocean depths from place to place. In fact this is of prime importance, seeing that the potential vorticity mentioned earlier is inversely proportional to the local depth. Hence, we find a whole class of motions often called 'topographic Rossby waves'. A map of the potential vorticity of the oceans was compiled by Peter Rhines when he was a PhD student at Cambridge University. As a result, we find that it is not only gravity waves that can be guided or trapped by bottom topography (the trapping of gravity waves by Macquarie Island southwest of Australia was explained in this way by me in 1967)[28] but also planetary waves and waves influenced by both gravity and the Earth's rotation. Many such ideal cases were modelled. In the complex situation of the real ocean, modern digital calculations are necessary, as in the case of the tides.[29] Nevertheless, idealised models have been and still are useful for interpreting numerical calculations and for suggesting certain parts of the ocean for special exploration. A good example is that of 'shelf waves' predicted to travel northwards along the continental shelf west of the British Isles. These were later detected by Cartwright and his colleagues.[30]

Wave statistics

A knowledge of the probability distribution of wave heights is particularly useful for the design of oil rigs and other offshore structures subject to wave action. For example, since high waves may destroy the platform of an oil rig, and every foot added to the height of a rig may add millions of pounds to the cost, the successful prediction of wave heights in a given sea state is important economically.

With such applications in mind I developed a theory for the statistical properties of random surfaces in both one and two dimensions, in terms of their energy spectra. In a series of papers in 1957,[31] I extended and generalised the classical analysis of Rice (1944-45) for noise in an electrical signal so as to find the statistical properties of a random, moving surface.[32] When Henry Charnock, after whom the Charnock constant is named, remarked to me that the sea surface did not really "look the same upside-down" since the wave crests were sharper than the troughs, I took up the challenge and produced a nonlinear statistical theory for relatively steep waves.[33] Among other applications, this has been used for calculating a correction to the sea surface level as measured by radar altimeters, from which, in turn, ocean currents may be calculated.

Breaking waves

The statistical theory relied on the assumption that the sea surface was at most only slightly nonlinear; the surface slopes were not too large. Wave breaking, on the other hand, is a strongly nonlinear phenomenon. Breaking waves and whitecaps in the deep ocean are both interesting and important. They strongly affect the transfer of gases between ocean and atmosphere–gases such as oxygen and the greenhouse gas carbon dioxide. After returning from the USA to Britain in 1969, I embarked on the serious study of breaking waves, using all the theoretical and experimental means available, and inventing others.

In 1969, I had been appointed to a Royal Society Research professorship at Cambridge and at NIO, so that I could do theoretical and experimental work at Cambridge and keep in touch with field observations while at NIO. I commuted at least once a week. With Mark Donelan, a Canadian post-doc who came to work with me at Cambridge, we studied the breaking waves generated by a motor launch travelling parallel to a river bank (the river Cam). In the ship-towing tank at Feltham, Middlesex, we studied the turbulence produced below the crest of a breaking wave. And we showed how to produce a breaking wave in a laboratory wave channel (in Cambridge) by generating first short waves and following these up by gradually increasing the period of the wavemaker. In this way the wave energy could be made to converge at one point at a certain distance from the wavemaker, and so to produce a plunging breaker.[34] The experimental method has since been used to great advantage by Ken Melville and his students, first at the Massachusetts Institute of Technology (MIT) and then at Scripps Institution, in their laboratory studies of breaking waves. The work continued in the new wave tank built at Wormley in 1966.

Stewart Turner, another member of DAMTP, had noticed how, in a flight over the ocean between Perth and Adelaide, the 'whitecaps' appeared and disappeared periodically with a period just half of the wave period. Together with Donelan, who studied the same phenomenon from a ship on his way back across the Atlantic, we wrote a letter to *Nature* explaining this phenomenon.[35] It is due to the fact that the steepest waves in a wave group break only near the middle of the group, and the group travels with only one half of the phase speed. Thus if you watch waves approaching you from a platform or a pier in deep water, a breaking wave will most likely disappear before it reaches you! Another wave from twice as far away will be the one that breaks where you stand.

I also devised some simple but useful expressions for travelling waves and for standing waves of limiting height; the travelling waves have a sharp angle of 120° at their crests, as first shown by Stokes (1880). Spilling breakers, i.e. those with a region of white water on their forward face, were modelled by Turner and me theoretically, and we explained a pronounced intermittency in spilling breakers first noticed in a movie of laboratory waves taken by S.P. Kjeldsen at Trondheim in Norway.[36] In succeeding years, studies of breaking waves were to be greatly developed.

The Cambridge connection

One very notable feature of wave studies at NIO, extending from the time of Group W onwards, was the intimate collaboration between Cambridge mathematicians and the physical oceanographers at NIO. This applies especially to the upper ocean. Fritz Ursell and I were both young mathematicians from Cambridge, as was also Jim Crease, who joined in 1949. John Swallow had taken Part I of the Maths Tripos, but then transferred to Physics. Owen Phillips was a research student in DAMTP before becoming a Research Fellow at St John's College. Ursell and I both were research Fellows at Trinity, the College of G.I. Taylor and G.K. Batchelor. Ursell was later a lecturer in DAMTP before becoming a professor at Manchester University. Steve Thorpe, Phillips' research student at Cambridge, was also at Trinity College.

The 6-monthly series of seminars held alternately at NIO and at DAMTP were highly significant. A party of perhaps a dozen would travel from Cambridge to Wormley or *vice versa*. These seminars consisted of several presentations of new material, either field observations or theory or both. Thus Phillips and Brooke Benjamin both presented their original ideas at these meetings, and outside speakers included both Sir Geoffrey (G.I.) Taylor and Sir James Lighthill, who was drawn into oceanography during his 5-years' directorship of the Royal Aircraft Establishment (RAE) at Farnborough (1959-64). Particularly memorable were Lighthill's physical explanation of Miles's theory of wave generation by a turbulent shear-flow in the air, and his treatment of waves in nonlinear dispersive systems. Lighthill later became a member of NERC, and from 1965-70 was Chairman of its Oceanography and Fisheries Committee. A contemporary of Ursell as a Cambridge undergraduate, in 1969 he became Lucasian Professor of Mathematics in the University, a successor to Sir Isaac Newton.

The value and success of this regular seminar series is doubtless due to the fact that ocean waves and currents are mathematically well-behaved; mathematics is an essential tool for their elucidation. Secondly, Cambridge had, and has, a long tradition in applied mathematics, not least as applied to the ocean and atmosphere, from Isaac Newton, George Airy, G.G. Stokes, Lord Rayleigh, Horace Lamb, G.R. Goldsbrough, Harold Jeffreys and G.I. Taylor.

12

Waves, surges and tides

David Cartwright

Experiments with the Shipborne Wave Recorder

I was recruited by the Royal Naval Scientific Service (RNSS) in 1951, with a mathematics degree and a love of the sea, but I was assigned to a branch of the Admiralty near Bath known as DNC (Department of Naval Construction), which designed the construction of naval ships. I turned out to be a misfit in the hierarchy of DNC, and after two years I started desperately to look for other openings within the RNSS. I was particularly attracted by the *Reports on Wave Research* issued by Group W of ARL, and saw ready applications of their ideas and methods to outstanding problems in the area of ship motions which concerned DNC. However, the terms of my post did not include the experimental research needed to pursue such ideas. Essential to the scheme I had in mind were the use of Tucker's 'Shipborne Wave Recorder' (SBWR), described in Chapter 16, and Barber's spectral analyser, together with a host of statistical techniques for wave-like motions, explored notably by Michael Longuet-Higgins.

In 1953, I contrived to visit Group W at Teddington and to join a short sea trip off Plymouth to see the SBWR in operation. Deacon and Tucker were most welcoming and I immediately felt an affinity with all the researchers I saw there. Here, at last, was a sympathetic group of scientists to whose work I felt I could contribute. After negotiations between Deacon and the RNSS, I was seconded to the NIO at Wormley as from June 1954. I later learnt that Deacon had a reputation for taking on staff who did not fit into conventional slots, and making marine scientists of them. For my part, June 1954 marked a very significant turning point in my career.

Deacon suggested that my background was best suited to a study of the papers of Longuet-Higgins, who was just about to return from two years at Cambridge, with a view to assisting him in the practical application of his theories. This proved productive. I also learnt a lot about modern instrumental techniques in oceanography from Tucker. Longuet-Higgins had suggested that the SBWR should be tested by recording waves with the ship underway on a sequence of 12 evenly-spaced courses, each held for about 20 minutes. This manoeuvre became popularly known as a 'threepenny bit', from the shape of a current coin. The idea was that waves from different directions would be separated by spectral analysis on account of their different quasi-Doppler shifts. One 'threepenny bit' had recently been executed, and Deacon suggested I should apply the spectral analyser to its analogue traces. The results neatly showed a broad spectral peak of a period 8 secs from the east, modulated sinusoidally with wave direction relative to the ship, and a superposed smaller peak of a period 15 secs propagating from the south. The latter peak indicated a superposed swell from a remote source. Later, I applied the same process to records of ship motion.

A more sensitive instrument for monitoring wave directions was the NIO pitch-and-roll buoy which recorded wave height through a vertically mounted accelerometer together with the two components of wave slope; effectively the heave, pitch and roll of the free-floating buoy. Such an instrument had been suggested by Barber in the mid 1940s, but its construction was only completed in 1955, by which time a microbarograph sensor had been added to monitor air pressure at the surface of the waves. At that time practically nothing was known about the spread of directions in wind-generated waves or about their generating pressures: existing theories about wave generation were speculative and controversial. Our deployments of the pitch-and-roll buoy from *Discovery II* in 1955-56 resulted in a set of 16 records, each about 20 minutes long, of which five were selected as representative of local wind generation and relatively free from faults. Digitisation of the initial photographic traces of wave height, slope and wind pressure onto punched cards preceded our first cross-spectral analysis by digital computers, which had been put at our disposal by the nearby Royal Aircraft Establishment.

The results were the first of their kind and quite outstanding. We showed that the waves and their surface slopes had Gaussian statistical distribution, and that their spectra tended to conform at high frequency to Phillips' inverse 5th power law. We derived useful numerical expressions for the spread of wave direction generated by real winds. Air pressure at the surface was found to be largely in anti-phase with the height profile, as expected, but with a slight phase-lag accounting for the generating process. We compared the results with a recent theory of Philips for the initial stages of wave-generation and with a theory due to J.W. Miles for the later stages where instability takes over. For the waves we encountered there was evidence that the exponential growth rate postulated by Miles was dominant.

In May 1961, a four-day Conference on 'Ocean Wave Spectra' was convened by the US National Oceanographic Office and the US National Academy of Sciences at Easton, Maryland. All of Deacon's scientists engaged in wave research were invited, as well as other leaders in the field such as Hasselmann, Munk, and Pierson. Even Barber showed up from New Zealand. Deacon was invited to give a keynote address, and our wave-buoy work was presented by Longuet-Higgins to great acclaim.[1] I presented a paper of my own on a related subject.

With the help of Norman Smith of NIO, I later experimented with two other types of wave-measuring buoys, designed for improved directional resolution. The first was a 120 m line of simplified pitch-and-roll buoys, to be towed as slowly as possible by the ship. The second was the 'Cloverleaf Buoy' which estimated the three components of curvature of the wave surface as well as the two components of slope. The results of experiments at sea using these buoys were interesting, but both systems proved clumsy to handle and so presented little effective advance on the original pitch-and-roll buoy itself.[2] The pitch-and-roll buoy was subject to intermittent jerks from its towing cable, causing sporadic changes in its 'pitch' signals. The capability of the cloverleaf buoy to resolve wave directions was limited to a rather narrow band of wavelengths, but was later put to good use during ship trials by Norman Smith and John Ewing, a late recruit to the NIO staff. The technology was later taken over successfully by Japanese wave researchers.[3]

During the 1960s, oceanographers began to realise that a single research ship was insufficient for modern analysis of systems such as waves: what were needed were simultaneous measurements by a co-ordinated group of well-separated research vessels, or later, of bottom-moored instruments. Such schemes usually required international collaboration.

Walter Munk had organised a line of swell recorders along a great-circle across the Pacific Ocean from New Zealand to Alaska, and much work had been done compiling the statistics of wave spectra and correlating them with the local windfield–e.g. by Darbyshire of NIO, and by Pierson and Moskowitz of New York. Deeper understanding of wave generation required a line of recording stations spaced along the 'fetch' of the wind. First to achieve this was Klaus Hasselmann of the Max Planck Institut für Meteorologie at Hamburg. With assistance from the Deutsche Hydrographische Institut, also at Hamburg, the Netherlands Meteorological Institute (KNMI), the Westinghouse Research Laboratory at San Diego, and NIO, a 160 km line of 13 recording stations was maintained in the North Sea, west of the German island of Sylt, during July 1969. This exercise became known as The Joint North Sea Wave Project (JONSWAP). The line comprised five ships and eight bottom-mounted recorders. NIO pitch-and-roll buoys were operated by scientists on every ship. Currents were also recorded where possible.

Hasselmann's main objective was to demonstrate the importance of energy transfer within the directional energy spectrum by nonlinear resonant interaction. This rather complicated physical process is separate from the downward transfer from the wind to the sea waves, and it results in energy depletion in some zones of the directional energy spectrum and energy growth in other directions. Among other features, it accounts for the steady growth at the long-wave (low-frequency) side of the spectral peak, which is hard to explain by elementary physics. The concept of nonlinear resonant interactions had been discussed theoretically at the Easton Conference in 1961, but the rate of energy transfer involves massive integral computations and so were far from practical evaluation, or even belief in their reality by some.

Not all the wave recorders used during the JONSWAP exercise worked successfully, but the experiment as a whole can stand as the first demonstration of nonlinear interaction in sea waves and the realisation of a true source-function for the growth of a wave-spectrum in a natural wind.[4] A parameterisation of the JONSWAP spectrum into its salient characteristics of peak frequency and spread has proved to be accurate and widely applicable to the statistics of independent wave measurements. A separate objective of JONSWAP, to trace the action of bottom-friction on swell waves, proved indeterminate.

The motion of ships at sea.

Going back to 1955, Deacon was keen for me to carry out experiments on the oscillatory motions of *Discovery II* in waves, using the facilities developed at ARL and NIO. In the mid 1950s, the only tool available for testing a ship's sea-keeping qualities was by towing a reduced-scale model in periodic waves generated in a long tank. Here was an opportunity to show not only how modern wave-measuring equipment could extend such tests to full scale at sea with waves coming from any direction (i.e. not only in head seas), but also to introduce modern ideas for analysing random wave-like motions and their statistical properties. For principles of naval architecture, I obtained valuable collaboration with Assistant Professor Louis Rydill at the Royal Naval College which was then at Greenwich.

After I had completed further measurements at sea based on 'threepenny bit' manoeuvres, Rydill and I worked up a paper on the pitching and rolling of a ship at sea which we presented to the Royal Institution of Naval Architects in 1957.[5] A second paper, concentrating on the vertical oscillations alone, was presented by me in 1958 at a Symposium at Wageningen in the Netherlands.[6]

These papers attracted the attention of Dr F.H. Todd, Superintendent of the Ship Division of the NPL, who wished to organise new approaches to ship research in the UK. The outcome of Dr Todd's initiative was an agreement for a new programme of sea-keeping experiments by NIO, NPL and the British Shipbuilding Research Association (BSRA). The first ship tested was a 'Weather Ship' (a converted naval corvette) which occupied a station 'Juliet' west of Biscay in 1959, by arrangement with the Meteorological Office.[7] Tests on other ships of more commercial importance approved by BSRA, followed in the early 1960s. Deacon was pleased to see NIO playing a useful advisory role in the national world of engineering.

Sea level, tides and storm-surges

From the mid 1950s, a research programme of national importance was engendered by the severe sea floods in winter 1953 along the coasts of the southern North Sea including London and the Thames estuary which killed about 300 people in the UK and nearly 2,000 in the Netherlands. In Britain, the principal research organisation involved in such work was the long-established University of Liverpool Tidal Institute (ULTI), and a steering committee was set up by the Ministry of Agriculture, chaired by Joseph Proudman, who had also been the founder of ULTI. The Meteorological Office, NIO, and the Ordnance Survey sent representatives and participated in the work.

Flooding at Sea Palling, Norfolk during the North Sea storm surge in January 1953. Copyright Archant Norfolk

Although NIO eventually took part in many aspects of the work promoted by the 'Flood Committee', as it was usually called, its early role was merely to examine the external surge emanating from the Atlantic Ocean. To this end, Jim Crease set up new tide-gauge installations in the Outer Hebrides and Shetland, and he constructed a mathematical model of a Kelvin Wave diffracted round a land barrier into a semi-enclosed sea.

In 1959, Dr Deacon invited a Japanese electronic engineer, Dr S. Ishiguro, to work on the response of shallow seas to air pressure and wind-stress by using networks of electrical analogue circuits to simulate the generation of a storm surge, then quite a novel approach. Suzhen (we called him Shichan, the diminutive used by his family) Ishiguro came to us from the Nagasaki Marine Observatory on a UNESCO fellowship in 1956 and, after a short return visit to Japan, Deacon offered him a job at NIO on the permanent staff.

In the electronic analogue model of the North Sea the inertia of the water was represented by inductors, the storage capacity of the surface area by capacitors, the bottom friction by resistors, and the Coriolis force due to the rotation of the Earth by rather more complicated circuits. The North Sea was divided into a grid, and each grid cell contained these components plus arrangements for feeding in the wind stress and atmospheric pressure. The grids were, of course, interconnected, and the water depth in each had to be taken into account.[8] The best summary of Ishiguro's achievements is contained in a general survey he wrote of analogue methods in oceanography.[9] But he also could take pride in his son, Kazuo, who became one of Britain's top novelists, winning both the Whitbread and the Booker prizes.

The North Sea was obviously of most interest to the Flood Committee and it became the focus of most of Ishiguro's research. He developed a series of North Sea models over many years with successive refinements as new electronic components became available, but their use as a practical working tool was eventually overtaken by digital models developed by Norman Heaps and Roger Flather at the University of Liverpool Tidal Institute (ULTI).

As a side-issue from North Sea modelling, Crease responded to a need of the Ordnance Survey for geodetic levelling between England and France by comparing the mean sea-levels measured by tide gauges along the Channel coasts, with a correction for the transverse slope caused by the action of the Coriolis stress on the mean current along the Channel. This correction could be estimated by the difference of electrical potential along an underwater telephone cable, of which NIO had first-hand experience through the work of Longuet-Higgins and Bowden. Crease and I later applied the levelling exercise to two years of simultaneous tide gauge data at Ramsgate and Dunkerque with the help of British and French authorities.[10] We found that the slope correction was 79 mm up towards France. The geodetic datum at Dunkerque turned out to be 196 mm higher than the Ordnance datum at Ramsgate. The difference in datums was eventually confirmed by the Ordnance Survey by direct geodetic levelling through the Channel Tunnel.

Tide predictions by computer

Working with tidal data inevitably introduced us to the traditional lore of tide prediction, for which ULTI was the national authority. Until about 1960, ULTI was committed to the use of an elaborate mechanical device for its tide predictions. We found it more convenient to produce our own predictions by digital computer, then still a novelty for many scientists in the UK, especially in the field of tides. In 1963-65, I was invited to work with Munk at Scripps who was developing a modern approach to the analysis of tides, made possible by the large electronic computers available in the USA. The analytical scheme which Munk and I worked up was based on the physical response of the ocean to the lunar and solar tidal forces, and became known as the *Response Method* as distinct from the traditional *Harmonic Method* with its roots

David Cartwright inspecting output from the IBM 1800 computer at Wormley.

in the late 19th century. Tide predictions by the *Response Method* were demonstrably more accurate. Our paper[11] was acclaimed as a 'landmark' in the literature on tides, though in practice use of the *Response Method* was limited to a few experts.

When a member of his staff wanted to pursue a promising idea in marine science, Deacon tended to adopt a *laissez-faire* policy. On my return to NIO after 17 months in California, Deacon allowed me a free hand to apply my newly acquired knowledge to the aims of the national Flood Committee. NIO had yet to acquire its own mainframe computer, but an arrangement was made for NIO staff to use the (then) 'large' computer at IBM in London. On one occasion, Longuet-Higgins had consumed such a heavy load of computer-time in computing a series of eigen-functions that IBM allowed us 5-6 hours of computer time without charge. For my part, I was able to construct a forecasting system for hourly sea levels around the British North Sea coast, using a *Response Method* not only for the tides but also for the surge-response to pressure and wind and their dynamic nonlinear interactions between tide and surge.[12]

Recording offshore tides

Another export from Munk welcomed by Deacon was a project to develop instrumentation for recording the tide in the open sea through changes in pressure at the seabed. The idea was not new, but very high precision was required to record changes of a few millibars against the ambient pressure of at least a few hundred to a few thousand bars. The instrumental technology was only just becoming possible for such high precision. There was then a dearth of knowledge about the tides offshore which, if well known, would have important applications to geophysics and lunar theory as well as offshore engineering technology. Attempts by mathematicians since Newton's time to solve the dynamic equations of the global tides had failed: existing measurements at harbour tide gauges were too distorted by coastal topography to be extrapolated far offshore. Once the technology had been established, Munk envisaged a network of oceanic tidal pressure measurements which could be used as the basis for digital models. In this he had the backing of UNESCO's Scientific Committee on Oceanic Research (SCOR) which was willing to provide money to aid workshops and travel. In 1965, only the US and France had realistic plans to participate in such a scheme. We at NIO were able to join them shortly on behalf of the UK.

Central to the overall problems was the global rate of dissipation of tidal energy, then (in the 1970s) thought to be in the region of 2 Terawatts (a Terawatt is a million, million watts) for the lunar half-daily tide alone. 2 TW is quite a small quantity compared with the amount of energy already raised in the ocean tides, but quantification was vital to the understanding of, for example, the steady increase in the length of the day, the rate of recession of the moon from the earth, and, as it later turned out, the capacity of the ocean circulation to overturn.

Short of a mathematical solution, a solution for the mean rate of dissipation could in principle be obtained by estimating the rate of working of the (known) tidal stresses on the ocean, and its divergence (method 1), but this would require measurement of pressure variation at every point of the earth's surface. Alternatively (method 2) one could divide the area into finite seas with tidal measurements along every boundary, including tidal currents. The currents are in principle depth-averaged, so more difficult to estimate. Method 2 was thought to be more feasible at the start of the SCOR-funded campaign, on the assumption that all dissipation occurs in shallow seas due to bottom friction. That assumption was called into question as dissipation by internal tides in the deep ocean became quantifiable, and indeed internal tides are now known to constitute a non-negligible sink of energy.

For this campaign, NIO had a 'laboratory' at hand in the complex of shelf seas surrounding Britain and Ireland, mostly washed by large tides. As later technology developed, we also extended external boundaries across the deep ocean, from Portugal to Canada, from Iceland to the Azores, and between West Africa and North Brazil.

NIO scientists were familiar with the principles of pressure recording from their work on waves, but there was a long way to go before reliable and accurate–tenth of a millibar–records could be made in even 200 m depth. Temperature sensitivity was a bugbear because temperature also varies tidally, and ability to record for at least a month before recovery of the instrument was another challenge. Acoustic release systems were still in their infancy. It was necessary to record continuously for at least a month in a depth of at least 150 m, ideally in 2,000 m or more, with precision better than one part per million. Tucker, then head of the Applied Physics Group, provided the initial lead with capacitance-plate pressure sensors and digital recording. The techniques were developed by Peter Collar and Robert Spencer, who eventually exceeded the desired goals by the use of refined technology. Collar's and Spencer's first experimental deployment in 1969 in the Hurd Deep west of Guernsey recorded for 14 days at 104 m depth and proved the technology to be viable; by 1971 the team was getting excellent month-long records at the shelf-edge in 188 m.[13]

Our first plan was to record tidal variations of pressure along the shelf-edge surrounding Britain and Ireland between Norway and Brittany at intervals of about 50 sea miles, with the object of providing tidal boundary conditions for a relatively large expanse of shelf-seas bounding the northeast Atlantic. Where feasible, the pressure data were supplemented

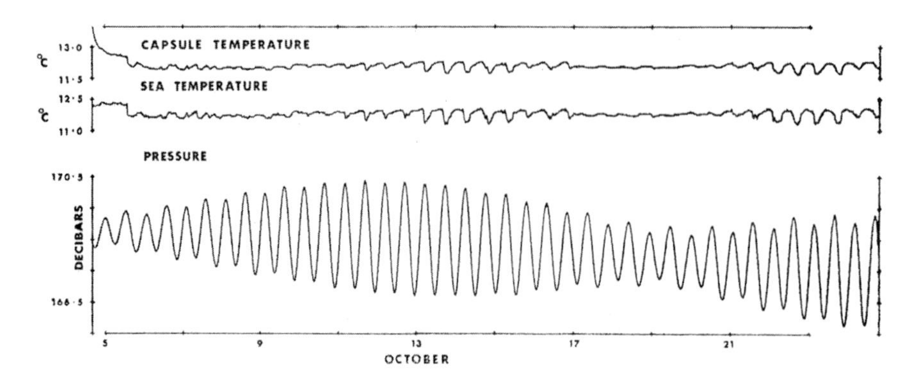

One of the early offshore tide records from La Chappelle Bank in October 1969. Note the 14 day springs-neaps cycle and also the downward drift in the average pressure due to sensor drift.[13] *(1 decibar is approximately 1 metre.)*

by currents. In the region south of the island of St. Kilda these revealed an area of enhanced diurnal (changing once per day) currents, not obvious from the pressures alone. A similar effect had been noted for centuries by fishermen in the Sound of Harris (Outer Hebrides), but never satisfactorily explained. Our measurements showed for the first time that the diurnal currents were spread over a relatively wide area of shelf and could be explained by a form of resonance known as shelf waves associated with the local bathymetry including the deep ocean.

The programme together with careful analysis by the *Response Method* took until about 1978 to complete, by which time our technicians had extended the depth capability to 3,000 m with the use of Digiquartz sensors. However, this takes us into the era of the IOS.

An international calibration exercise

The last event in the NIO tidal programme before its transfer to IOS at Bidston, near Liverpool, was a calibration exercise for pressure recorders deployed from *Discovery* towards the end of 1973. All active members of the SCOR Working Group 27 on 'Tides of the Open Sea' were invited to compare their pressure recording techniques for a month at two sites near the shelf-edge west of Brittany. One of the recording sites was fairly deep at about 2,000 m; the other was a 'shallow' site at about 200 m, both in reasonably flat bathymetry. Thus, both types of technology, shallow and deep, were catered for. Participants included the US, Canada, France and the UK. The UK was the only team to feature instruments for both shallow and deep technology. There were a few unfortunate losses of equipment at both sites, but the instruments recovered showed remarkable agreement in tidal amplitude and phase, and were therefore worthy of continuing the programme.[14]

Forward look – the impact of space geodesy

On a clear evening in October 1957, many staff of NIO, including Deacon, assembled on the flat roof of the NIO building to witness a passage of the first Russian satellite, *Sputnik 1*. Though vastly impressive, it was not immediately apparent how the new technology could benefit oceanography. But in the ensuing years infrared and microwave sensors mounted on satellites slowly became accepted as useful tools for monitoring the ocean, opening up radically new approaches to the subject. The study of ocean tides, as well as much dynamical oceanography, was transformed when satellite radar altimeters backed up by high precision space geodesy began to monitor heights of most of the ocean surface to precisions of at first a few decimetres, then steadily refining to as little as a few millimetres by the 1990s. Some physical oceanographers (including myself) started to follow developments from the later years of NIO onwards, and we played a palpable role in the choice of suitable orbits at meetings of the steering committees of NASA and of the European Space Agency. The success stories of the US/French TOPEX-Poseidon and the European ERS satellites and their sequels owe a lot to research into ocean tides which started in the sea-going efforts of SCOR's tidal Working Group, as well as other disciplines in physical oceanography.[15]

13

Applied wave research

'Tom' Tucker

The research on wave theory by Longuet-Higgins and Cartwright and the development of wave prediction methods by Darbyshire underpinned the application of new concepts to the solution of very practical problems that were brought into sharp focus with the development of oil and gas fields in the North Sea. They had enormous economic consequences. In 1996[1] Laurie Draper, my colleague at NIO for many years, stated that "*In the eighties, a submission by a distinguished engineer with no direct involvement in the Institute (by that time this meant IOS but it could equally well apply to NIO) stated to a House of Lords' Select Committee on Marine Science that in his opinion the Institute's wave research had saved £1,000 million and probably several hundred lives in the North Sea alone*". That saving was a return on an investment in NIO/IOS wave research that at present-day prices would amount to £50 million.

Draper was a key player in applied wave research. He joined the Applied Physics Group in 1953 and after a brief spell working on instruments he devoted the rest of his career to working on waves. His main interest was the application of our knowledge and understanding of waves, and he became our main interface with the engineering profession. His contribution to this field was acknowledged in 1965, when he was awarded the Telford Premium Prize by the Institution of Civil Engineers, and just before he retired, when the Society for Underwater Technology awarded him their Ministry of Defence Award for Oceanography. He was invited to set up the Canadian Wave Climate Study, and in 1968, was appointed as its initial Director for a year with a funding of C$ 2 million per annum and a staff of 20.

As explained by Fritz Ursell in Chapter 4, Jack Darbyshire was given the job of improving the methods for forecasting the waves in a local storm. Fritz also points out that forecasting the waves for the D-Day landings was relatively simple because the inner English Channel is, in effect, an inland sea, so Jack's task was to produce more accurate and more generally applicable formulae.

If one considers a notional situation in which a strong wind is suddenly 'switched on' blowing north-to-south across the English Channel, then if one sat on a ship in the middle of the Channel one would first of all see a short steep sea and the waves would get steadily longer and higher with the increasing duration of the wind until an equilibrium is reached. If, after this, one sailed from the north shore southwards, one would again experience a short steep sea at first and the waves would get longer and higher as the fetch increased downwind. Thus, the important meteorological parameters in wave forecasting before the days of huge numerical models on computers were wind speed, direction, fetch and duration. These were calculated from the usual meteorological charts using the spacing and direction of the isobars. Of course, in practice it was rarely as simple as this makes it sound, and Jack's job was to try to improve the practical methods. At that time, we had very little understanding of the physical processes involved in wave generation, so the formulae used were largely empirical. This was one of the reasons that we were under constant pressure to provide more and better wave data. By 1952, he had developed empirical relationships between the wind conditions and various wave parameters.[2] These enabled him to predict waves caused by local storms and by the swell from distant sources. However, the results and methods were based on data taken with our shore-based wave recorders, and initially we did not realise how much storm waves were attenuated in what seemed to us to be quite deep water (15 to 20 m). In fact, we did not properly understand the processes until after 1973.

Instrumental wave data from the deep ocean started to be available when the first Shipborne Wave Recorder was installed in one of the Meteorological Office's Ocean Weather ships in 1953. The ship occupied alternately one of two stations (India–61° 00'N 15° 00'W and Juliet 52° 30'N 20° 00'W) in the NE Atlantic. The waves there were much higher than near shore. Thus by 1955 Darbyshire had developed a different set of criteria for forecasting them.[3]

Darbyshire continued to develop his prediction methods as more and more data became available, but engineers found them difficult to apply in practice, so in 1963 Draper persuaded Mollie Darbyshire (Jack's wife) to work with him to produce a set of prediction graphs based on Jack's results.[4] Draper claims that this must be the most reproduced paper ever published by NIO or IOS, running into tens of thousands of copies and re-published many times worldwide (including translations, for example, into Japanese) and appearing in numerous books and articles.

In 1961, Darbyshire was appointed Visiting Professor of Oceanography at the University of Cape Town, and soon after he came home in 1963, he was appointed to the new Chair of Physical Oceanography at Bangor, so he and Mollie had little time with us after 1961. Draper effectively took charge of the analysis and interpretation of wave data, as well as the supply and installation of wave recorders.

As soon as coastal engineers became aware of our developing understanding of waves, they started consulting us. An early example concerned wave energy conversion, but the seminal episode was in connection with a new harbour being built at Tema, in Ghana. We were first contacted in 1953, just before we left Teddington, by an engineer from Sir William Halcrow and Partners. They were the consulting engineers for the harbour and before they had completed the first harbour wall, it had been overtopped by waves higher than the extreme wave for which the harbour had been designed. They had to stop work and think again!

Since there were no instrumental wave data from that part of the world, they had sent a young engineer out there for a few weeks 'to sit on the beach to observe the waves and to talk to local fishermen'. Now the coast of Ghana is in the Doldrums, so there are few local storms of any significance. The largest waves would be the swell from storms 5,000 miles away in the South Atlantic. Using his prediction formulae, Darbyshire estimated the height of the highest wave likely in 50 years, and found that it was about twice the design height used by Halcrows. However, it was obviously not a very accurate value, and Halcrows accepted that they would need instrumental measurements. It was not practicable to use shore-connected instruments, so they used a self-contained pressure

Satellite view of Tema Harbour. It was the long south breakwater that had to be strengthened. (The picture is approximately 3.9 km wide and oriented north-south vertically).

recorder designed at the French Hydraulics Institute at Chatou. This had to be recovered every two weeks to replace the batteries and recording film, but eventually it gave satisfactory service, and two years' of data were analysed by Darbyshire.[5]

Following this episode, UK engineers realised the importance of getting good wave data before designing coastal engineering works, and they developed close relationships with us. Most of their consultations were not recorded, and only two resulted in published papers.

Throughout the last decade of NIO, we spent an increasing amount of effort on measuring and analysing waves so that in 1971 we were maintaining 20 wave recorders round the British Isles, and some 50 of our instruments were in use in other parts of the world. John Driver was mainly responsible for this installation and maintenance work, with the help of other staff as necessary.

North Sea oil and gas

The first British discovery of North Sea gas, in the West Sole field off the coast of Norfolk, was by the BP jack-up mobile drilling rig *Sea Gem* late in 1965. Only days later, on Boxing Day, this rig capsized while the legs were being raised in preparation for a move, and sadly, 13 lives were lost. Because UK engineers had no experience of offshore platform design, the fixed production platform for this field was designed by an American team. They had been highly successful in the generally quiet waters of the Gulf of Mexico and, I think, had got over-confident. Anyway, they did not enquire whether there were any wave data for the area. It soon became clear that they had greatly underestimated the design wave height. In the platform's second winter someone estimated the crest and trough heights of a huge wave by scaling them against the position of structural members of the platform–it exceeded the design wave height and we were at last consulted! By good luck we had a long series of data from the Ship-Borne Wave Recorder (SBWR) on the nearby Smith's Knoll Light Vessel, and could predict the extreme wave height from this. We had also developed methods for 'hindcasting' extreme waves using estimates of extreme winds produced by H.C. Shellard of the Meteorological Office and entering them into Mollie Darbyshire's graphs. Both methods agreed surprisingly well, and were within 3% of the presently accepted values for that area. The value was 25% higher than that used by the Americans.

Other platforms were soon built, but it was not long before one of these failed, although not catastrophically and no lives were lost. The main reason for the failure was not underestimation of extreme waves, but that the designers had not realised the significance of one of the main differences between the North Sea and the Gulf of Mexico. Whereas a

The Jack-up drilling rig Sea Gem *that collapsed on Boxing Day 1965. The cause of the failure was the embrittlement of the steel due to the cold water temperature.*

platform in the Gulf may experience a hurricane that is much more severe than the worst North Sea storm, the North Sea has many more moderately severe storms. These constant stresses produce serious fatigue, and it was this that caused the failure. It was the first such failure, so it is perhaps understandable that the design engineers did not take the possibility into account. The consequence for us was that it was not just extreme waves that we had to predict, but we also had to provide the statistics of every wave throughout the year.

The NERC Annual Report for 1972-73 states that

> *The activity of the Wave Data Section has been dominated by the offshore interests of the oil companies. Help has been given to numerous individual companies; to the UK Offshore Operators Committee in helping to formulate an environmental data collection programme and then participating in its implementation; to the Department of Trade and Industry by participation in their committees, such as the Offshore Installations Technical Advisory Committee, and in particular by calculating probable extreme wave conditions in British Waters, the results of which are now used as minimum standards in the certification of*

> *offshore structures. Technical advice has been given on a wide range of wave problems concerning, for example, the Portsmouth hovercraft disaster, numerous harbour design problems and ship loss legislation. There has been an increase in the number of seminars and lectures presented in the UK and abroad and the section has been particularly active on working groups and committees, both national and international.*

Applied wave research had come into its own.

Analysis of the wave data

Because it was not practicable to record continuously, we had to make two compromises when setting up a wave measuring programme. The first was on the length of record. If you sit on a beach and watch the waves, sometimes you get a series of low ones and sometimes a series of high ones. To get a good measure of the average wave height, you have to record for a considerable time. In the Group W days, we estimated that a 20 minute record was the optimum compromise. Similarly, one needs to take such a record sufficiently frequently to follow changes in the wave conditions. After initially taking records every two hours we finally decided that one record every three hours gave adequate accuracy and this became our standard. These early decisions stood the test of time, and were the norm until digital recording and analysis became cheap and easy enough to take longer and/or more frequent records.

Our recordings using ink on a paper roll had two advantages: the operator could see immediately if anything serious had gone wrong, and could take an interest in the results, which was important if you wanted conscientious record taking. We were receiving hundreds of records a week that had to be analysed manually.

When we started wave studies, the accepted measure of wave height, suited to visual observations, was the Significant Wave Height, defined as the average height of the highest one third of the waves. The symbol used originally was H_S (now $H_{1/3}$). The definition of 'height' was the difference between the height of a crest and that of the preceding (or succeeding) trough. The trouble was that in a stormy sea, there were little waves riding on the big ones, and these often produced a small trough half way up the main wave. So we changed the definition of a wave to be what happened between two successive up-crossings of the mean water level. The height of the wave was the difference between the highest and lowest levels in this interval. The wave period T_Z is the length of the wave record divided by the number of these waves in it. Measuring the average height of the highest one-third of these waves was a time-consuming process, and in the early days, Darbyshire used the height of the single highest wave in the record to characterise it. This was unsatisfactory because if, for example, you took a succession of, say, 10-minute records, the height of the highest wave in each record would vary randomly over a wide range. Some better method had to be found.

This was achieved as the result of two pieces of theory. The first was by Longuet-Higgins.[6] This paper was a landmark in wave research because it was the first to give a theoretical formula for the statistical distribution of wave heights. One of the important results was a formula for the expected height of the highest wave in a record and this will be referred to again below. However, for the present purpose we are interested in a different result, and to explain this, a simple piece of maths is necessary. If y is the instantaneous height of the water surface above mean water level, then mathematicians define a quantity m_0 as the mean of y^2. In the case of waves, this mean is the same if taken over space or time. Longuet-Higgins showed that $4\sqrt{m_0}$ is a reasonable approximation to $H_{1/3}$ as defined above. Now $H_{1/3}$ is an unsatisfactory measure of wave height because it cannot be handled theoretically, but m_0 is the basic measure used in virtually all wave theory. Thus, a quantity $H_{m0} = 4\sqrt{m_0}$ is now defined as the significant wave height, with the suffix m_0 used to distinguish it from the original definition.

The second key piece of theory was due to Cartwright.[7] His paper is entitled '*On estimating the mean energy of sea waves from the highest wave in a record*', by which he meant the highest crest. It was not immediately obvious that this work led to a satisfactory practical method for analysing wave records, partly because using just one wave would give a large statistical uncertainty in the result. The position of the mean level is also difficult to establish accurately on a graphic record. However, Cartwright also gave a formula for the expected height of the second highest wave in a record. I realised that by using this and the fact that the theory also applied to troughs, one could greatly improve the statistics. I showed that by measuring the heights of the highest and second highest crests, and the depths of the lowest and second lowest troughs in a record, all in relation to a mean line whose position is now not critical, and then applying Cartwright's results, one could get a good measure of H_{m0}.[8] Draper converted this method into a really practical routine.[9]

Cartwright's formulae were used in graphical form so that the whole procedure became simple and quick. This 'Tucker-Draper method' became the world-wide standard until convenient digital methods became available. Cartwright made no assumptions about the width of the wave spectrum, and using both crests and troughs removes the effect of second-order hydrodynamic nonlinearities. Thus, we had confidence in the accuracy of the method, and in fact several workers tested the method and found that it gave a reliable result whose statistical accuracy is nearly as good as that derived from calculating m_0 from every point in the record.

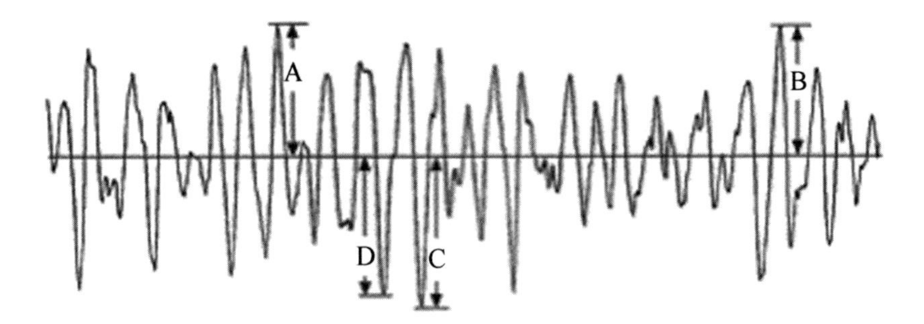

A wave record to illustrate the Tucker-Draper Method of analysis. In a real case, the record would be longer than shown here.

Using this technique, we acquired an extensive data bank of wave heights and periods from the mid-Atlantic and round the coast of the UK. This is unique in the world and has proved invaluable in many ways. For example, Carter and Draper,[10] using data from the Seven Stones light vessel off Lands End for the years 1962 to 1985, found an increase in annual mean H_{m0} of 36%! Subsequent studies have shown that this increase has not been maintained. The values are very variable but show a strong correlation with the North Atlantic Oscillation, as represented by the difference in annual average atmospheric pressure between Iceland and the Azores, for example.

Two types of information are required by the design engineers. The first is the height of the highest wave likely in a given number of years, which we shall take here as 50 years. The second is the number of waves in each height band that are likely to pass the structure in its lifetime. In both cases the period of the waves needs to be known.

For the highest wave, Draper developed a comparatively simple and logical method, which he used in the late 1960s. He started by using Longuet-Higgins' theory[6] to calculate the probable crest-to-trough height of the highest wave, symbolised by H_{max} (3 hour), in the 3-hour period centred on each record. This could be done using graphs of the theoretical relationships and the recorded values of H_{m0} and T_Z. (Michael later modified the formula using empirical data from real storm seas. He had assumed an ideal wave system in the 1952 paper.)

To obtain the extreme wave height, one calculates the proportion of the values of H_{max} (3 hour) that fall below a wave height h then plots this against h using appropriate axes and extrapolates the graph to the equivalent of one exceedance in 50 years. The extrapolation is way beyond the measured values and a lot of research went into the best way to do it.

For the statistics required for fatigue calculations, for each measurement of H_{m0} and T_Z one calculates the probable number of waves for each combination of height and period in the 3 hour interval. The conversion of these into the statistics required is straightforward in principle, but requires a computer.

A troublesome problem in the above is that there are always periods of time when the data are missing. If, for example, one were analysing one year's data and there were no data for January, clearly this would seriously bias the results. The simplest way of dealing with the problem is to 'fold in' the data from either side. In the case quoted, one would duplicate the data from the second half of December and from the first half of February. This should remove the bias, but the statistics are not so reliable as those using a full dataset. However, in some cases this simple method is not suitable and other approaches have to be used.

The JONSWAP spectrum

From the point of view of applied wave research, the most important product of the JONSWAP project (described by Cartwright in Chapter 12) was the JONSWAP spectrum. This was a universal spectrum for a wave system under active generation with the height and period parameters determined by the wind speed and fetch. It has been widely used in the design of wave forecasting systems and by offshore engineers, and has stood the test of time. In its original form, it applied to deep water, but a version applicable to finite depths and known as the TMA spectrum was later developed. The initials referred to the data sets used to derive it.[11]

Postscript

When IOS was formed in 1973, the wave data section was expanded and reorganised. The data banking and most of the advisory activities were transferred to the British Oceanographic Data Service (BODS) and the newly-formed Marine Information Advisory Service (MIAS). The research, instrumentation, data gathering and analysis, and data quality control were transferred to a new Engineering Oceanography Group headed by Ted Pitt at IOS, Taunton, where Tucker was in charge as an Assistant Director of IOS. Unfortunately, Draper could not move for family reasons, so was transferred to MIAS.

The Earth Beneath the Sea

The remit of the Director of NIO was to include the study of the earth below the sea. Arthur Stride, a geologist, showed that a development of naval sonar would be a valuable tool for exploring the continental shelf. Initial use produced a geological map and recognition of a pattern of sand transport and deposition unique to tidal seas. Anthony Laughton, a geophysicist with strong Cambridge connections and later Director of IOS, designed the first underwater camera in the UK for use in the deep ocean, compiled charts of the ocean floor and exploited the side-scan sonar which (as GLORIA) had been modified for use in the deep ocean. He initiated the UK membership of the Deep Sea Drilling Project. With these tools he and others in his group worked on the geological evolution of ocean basins, especially of the North Atlantic and the Gulf of Aden.

14

Side-scan sonar – a tool for seafloor geology

Arthur Stride

Beginnings

"*The National Institute of Oceanography has no obvious headquarters. It has little money and is not expected to last very long, so I advise you not to apply for the post of geologist,*" said the university's careers advisor. Nevertheless, I joined in 1950 but kept an alternative post on ice, just in case my adviser had been correct. The opportunity to work at NIO was especially attractive given the pressing need for information about the nature of the sediments on present-day continental shelves that would help geologists to recognise and interpret their supposed analogues in the marine sediments of the geological past. Although much was known about the sediments of accessible modern tidal flats, the sediments of the offshore tidal current environment were largely unknown. In the early 1950s, navigational charts were the main sources of information about the nature of the continental shelf, but they only provided widely separated spot depths and even scrappier information about seafloor composition. Indeed, compared with our knowledge of the land surface, it was very difficult to appreciate the nature of the seafloor in anything but a broad outline.

I started work under James (Jack) Carruthers, then based at the Hydrographic Department of the Admiralty in London. The initial task was to index new material that he was assembling for the future oceanographic library. This left time to read widely about oceanography, as well as to appreciate the scale of depth changes affecting sand banks in the North Sea, as revealed by Admiralty surveys. It was easy to see how the surveyors had defined the geographical limits of these sand banks from their depth soundings on navigational charts. But one could not understand the significance of small differences in depth between adjacent soundings, even when survey lines were close together.

A brief cruise on an American research vessel provided me with some cores from the Dogger Bank, a long, relatively shallow ridge stretching northeast to southwest across the middle of the North Sea. Its origin was uncertain. Seismic studies by American workers revealed that it consisted of material with low sound velocity, not rock. The cores showed that the uppermost materials were shelly sands and tidal flat deposits from the time of low Pleistocene sea levels. Fishermen had previously dredged up peat. We concluded that this large ridge was a moraine from the last glaciation that had become covered in vegetation until its surface was re-worked by the sea into tidal flats, before it was finally drowned when sea level rose to its present height. The discovery of a thick pile of sedimentary rocks under Dogger Bank was of great interest to oil geologists.

I subsequently spent two years at the Admiralty Underwater Detection Establishment at Portland, Dorset. Sonographs had been obtained with side-scan sonar. Attempts to interpret them by reference to scattered notations about seafloor composition on navigational charts were suggestive but inconclusive. Even so, the side-scan system could provide information about the seafloor in a band alongside the track of a ship out to a range of about 800 m. Portland scientists Derek Chesterman and Peter Clynick had produced a true plan-view mosaic of 25 km^2 of almost flat seafloor in Weymouth Bay. My task was to take samples from the seabed on a grid of points about 600 m apart, and to analyse grain sizes.

Divers inspected the seafloor at crucial locations. Grab samples and traverses with a towable underwater television camera showed that much of the floor was sand. There were also some areas of gravel, stones and molluscan shells in varying proportions, but there was only a small area of soft mud. We also took cores of about 20 cm in length. At some sites the sands overlay the coarser grades of sediments. For the study area as a whole, the size of the sediment particles was small compared with the wavelength of the sound used in the side-scan sonar equipment, so that individual grains of sediment could not be discriminated. However, each sediment particle at the surface of the seabed had the chance of scattering some sound back towards the ship, and altogether these tiny contributions provided a particular sound intensity (reverberation level) on the sonograph.

A uniform sand floor was represented by a relatively low reverberation level, whereas the rather varied coarser sediments were represented by a range of darker tones (higher reverberation levels) on the sonographs. All of these reverberation levels seemed to relate only to the surface of the seafloor and not to the coarser deposits buried only some 2 cm or more below the sand. There were too few samples to show whether the soft mud was effectively transparent.

These results were much more encouraging than the previous comparison in the North Sea, where sediment data had been obtained largely from navigational and fisheries charts and where the floor was later found to have much local compositional variation. Moreover, at some locations, undersea telecommunications cables had become so deeply buried by sand since they had been laid that they could not be lifted for repair.

I left Portland in 1954, convinced that side-scan sonar would prove to be a vital tool for exploring the seafloor surface. It was the only known method that could give a complete examination of a given area and could distinguish between sand and gravel. The scattered indications of these on navigational charts could now provide some of the ground-truth for interpreting sonographs.

The sonographs provided the third dimension that was missing from echosounder profiles, so that new types of information became available about the form and composition of the seafloor surface that gave fresh insight into its nature and origins. For example, some of the small differences in depth between adjacent soundings that had so puzzled me in the early 1950s were now seen as indicating parallel ridges or troughs of known orientation. Similarly, the method had already revealed the existence of sharply-defined, narrow, ribbon-like bodies of sand up to a few kilometres long in the North Sea aligned parallel to the path of the strongest tidal currents, as well as sand waves many metres high transverse to them. The sonographs had also revealed an additional bedform consisting of ridges and troughs extending parallel with peak tidal flows on the concave, eroding floor of one of the Edinburgh Channels that cut through a large sand bank in the seaward approaches to the Thames Estuary.

Secondly, side-scan sonar had revealed a familiar pattern that could only be interpreted as an area of eroded, gently sloping beds of sedimentary rock that had been displaced by a series of small faults. This clear pattern could have been mistaken for what one could see with one's own eyes on a neighbouring beach. This greatly increased confidence that the system might provide realistic information about other types of seafloor about which one had no previous knowledge. But many patterns on the sonographs were still uninterpretable.

Work with echosounders only

At Wormley, I was welcomed by George Deacon with the words "*I expect that you will find something interesting to do*". Such research freedom was most encouraging. Sadly, though, he was not persuaded that NIO should build its own side-scan sonar. Indeed, he considered that the seafloor patterns that had been discovered, though undoubtedly interesting, were too few in number and had been seen too locally to offer much hope that the new method would be of value as a general survey tool for the continental shelf. Accordingly, the Institute's hard-to-come-by money would continue to be spent on projects with a more certain chance of success. The disappointment could not even be mitigated by publishing the results of my work with Chesterman and Clynick, for this was not permitted by the Admiralty until 1958.[1] So, I had to find an alternative approach for exploring the continental shelf, while keeping the same goals in mind.

During the mid-1950s, knowledge of the sediments of the continental shelf gleaned from navigational charts had been supplemented by some much more representative seabed samples taken by French, Dutch, German and American geologists. Despite this, many of these sediments occurred patchily and were variable in grain size. They did not resemble the types of sediments found in sedimentary rocks, then thought to have been deposited on the continental shelves of the geological past. Researchers were dissatisfied with purely descriptive approaches to studying such supposedly present-day continental shelf sediments. It was deemed necessary to find undoubted modern deposits by gaining an understanding of the processes that were shaping these sediments on present-day shallow seafloors.

One needed to distinguish between modern marine deposits of the last 3,000 years (when sea levels were similar to those today), from older materials of the last ice age (such as the large glacial moraine of the Dogger Bank), as well as any sediments laid down during the 7,000 years while the sea level was rising to cover the beaches and rivers of the exposed continental shelf. Some of the most notable of the low sea-level marine features are aligned approximately perpendicular to the edge of the continental shelf to the west of Brittany. They are large, more-or-less parallel ridges, up to 110 km long, up to 15 km wide and up to 55 m high, with rounded cross-sectional profiles. Navigational charts showed that at the surface, at least, they were made of sand. Our seismic profiles revealed that their whole thickness consisted of sediment sitting on a floor of rock. They are larger than the numerous sand banks of the North Sea, but unlike those they are devoid of sand waves, as they are swept by much weaker tidal currents.

In the 1850s, English geologists had been aware of the importance of tidal currents for sediment transport near land. In 1925, Frenchman Louis Dangeard observed a rough correspondence between the grain size of the sediment and the strength of the associated tidal currents in the English Channel. In 1950, German Otto Pratje confirmed this and proposed that it applied more generally to the seafloors of northwest Europe. Dutchman Van Veen had suggested in 1935 that the steep northern faces of the large sand waves on the Dutch side of the Southern Bight of the North Sea indicated that they were being driven northward by the relatively stronger ebbing tidal current.

Denied use of side-scan sonar, I sought to explore the seafloor by means of echosounding. Echosounder records from the Lowestoft Fisheries Laboratory revealed the presence of sand waves up to about 8 m high on both sides of the Southern Bight of the North Sea. The relatively steep northern slopes of those on the Dutch side confirmed their inferred northward migration. In addition, the asymmetry of sand waves located seaward of East Anglia implied that they were migrating southwards towards the Thames Estuary Approaches. Here was another tantalising glimpse of an existing seafloor that was in an active stage of evolution.

Further afield, on the edge of the continental shelf west of Brittany we found some sand waves up to 7 m high, which were also surveyed by a biologist from the Marine Biological Association (MBA) in Plymouth. Because the oceanward face of the outermost sand wave led down on to the top of the continental slope, there must be occasions when a pulse of sand is discharged downwards. Thus, tidal currents affect the seafloor even at a depth of about 200 m, although in this case they are associated with tidal lee waves, when the east-going tidal current impinges on the top of the continental slope.

In 1957, money to make a side-scan sonar finally became available, albeit initially for fisheries research rather than geology. It was designed by 'Tom' Tucker and Ron Stubbs and made at Wormley. However, they were encouraged to modify the transducer so as to provide a narrow vertical fan-shaped beam that would be suitable for examining the seafloor.

Side-scan sonar survey begins

The equipment was fitted to *Discovery II* in 1958 and was used on a brief cruise during that spring. Fisheries research was done during the day, and the geological reconnaissance traverses were run at night, with the aim of demonstrating conclusively that significant geological results could be obtained both on a rock floor and on one made of sediments.

The first of these areas was on Lulworth Bank, located south of the Purbeck coast, where I had previously found some outcrops of sedimentary rocks. This revealed widespread stone-bands displaced by faults. A full survey would have to be deferred until more ship-time was available and surveying could be done during daylight hours, and with more accurate navigation.

The second area to be examined during that brief reconnaissance cruise was the bay that lies between Start Point and Lizard Point. Limited references to sediments on the navigational charts recorded the presence of sand, or else gravel, stones or shells. A biologist from the MBA had already found sand patches and it seemed possible that these might resemble those that we had found off the East Anglian coast in 1953. As predicted, the sonographs revealed the widespread presence of linear patches of sand. However, they were not as long, nor as narrow as those that had been found under the North Sea. Nevertheless, like those, they

Side-scan sonar installation on the hull of RRS Discovery II.

were also elongated parallel with the path of the strongest tidal flows and must have resulted from their action.

It was clear that side-scan sonar could provide new and interesting results. Although I was mainly interested in sediments, rock floors seemed likely to yield more immediately rewarding results. Rock outcrops would be recognised despite their variable appearance on sonographs, so they would be less likely to be confused with floors of sediment.

In 1959, the first detailed side-scan sonar survey of a floor of sedimentary rocks was made on Lulworth Bank, with the intention of producing a geological map. The upstanding stone bands were relatively easy to follow between adjacent sonographs, and the pattern of folding of the strata was evident. The rocks had been arched up into a large but simple asymmetric fold that represented the seawards extension of the Purbeck Anticline, as well as into some minor folds. This structure was found to be ruptured by many small but long, northerly-trending faults, with their downthrow side discernible in some cases. Neighbouring seafloor, further south and westwards into Weymouth Bay, was examined in less detail. Side-scan sonar had revealed the overall geological structure. Desmond Donovan of Bristol University determined the stratigraphic ages of rock samples. Later, Adrian Lloyd of University College, London (UCL), determined the micro-faunas.

Rock samples were taken on this cruise, supplemented by ones that had been taken in Weymouth Bay during 1952 and 1953. Subsequently, BP funded a small boat and divers from the Cambridge and Guildford Sub-aqua Clubs to work the inshore waters. Our geological map[2] was used by BP to site Britain's first offshore oil exploration borehole on

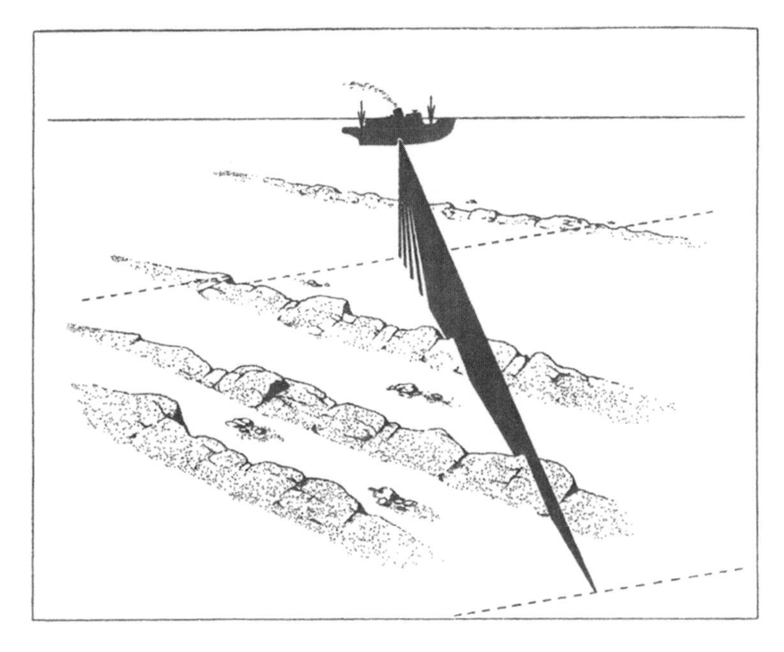

Schematic of visualistion of seabed features.

the core of the anticline seaward of their producing-field at Kimmeridge. Our survey also revealed that the portion of Kimmeridge Clay without stone-bands, that was swept by particularly strong tidal currents, had been deeply eroded. These local deeps, as well as the even deeper ones off Portland Bill, could not be accounted for by river-erosion at times of low Pleistocene sea level, and must have been made by sand-transporting tidal currents during a period of high sea level such as the present day.

The captain of a merchant vessel drew my attention to an isolated area of particularly rough seafloor about 100 km west-northwest of the Scilly Isles. Side-scan sonar on *Discovery II* revealed an approximately circular area of rough ground about 15 km wide, with peaks reaching 64 m high. This rocky ground was unlike any other rock outcrop pattern previously seen on the sonographs. Dredge samples showed that it was granite, rather than a sedimentary rock. A potassium-argon date showed that it was the westernmost member[3] of the suite of granites that outcrop along the spine of south-west England from Dartmoor to Bodmin Moor, St. Austell, Carnmenellis, Land's End and the Scilly Isles. Mining geologists wondered whether it, too, was rich in ores.

From then on, every opportunity was taken to use the equipment when our vessel was passing over the continental shelf. These reconnaissance traverses would ultimately extend from as far afield as Gibraltar, as well as north, west and east of the British Isles. Collaborative work with geologists from Bristol University and UCL, sought to find out what lay

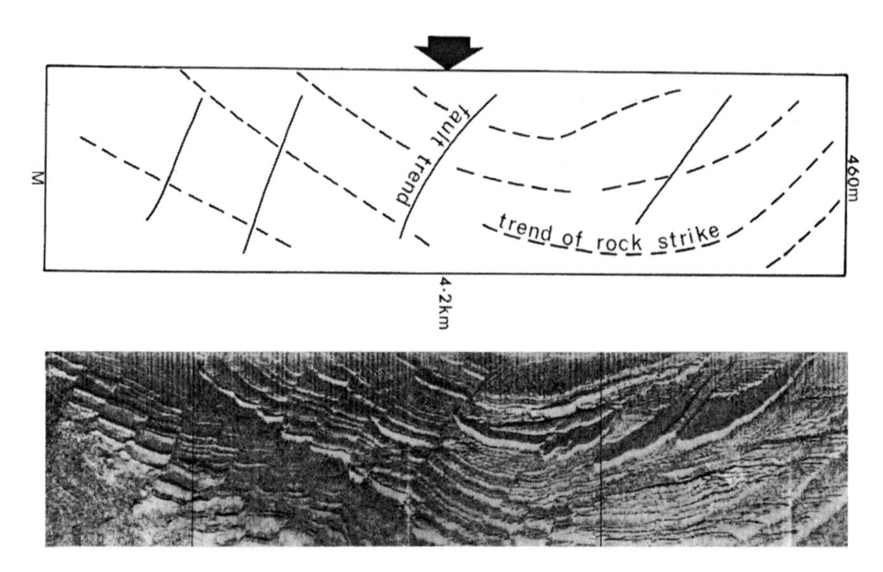

Curved outcrops of stratified rocks displaced by small faults on the floor of the English Channel revealed by side-scan sonar (4.2 km by 460 m). Arrow marks direction of insonification.

beneath the Bristol Channel. Its straight southern side makes this one of the most striking indentations of the British coastline, and various different origins had been proposed. Side-scan sonar quickly mapped the numerous stone-bands that had been etched out by erosion of the intervening softer clay layers.

Seismic reflection profiles revealed something of the underlying structure. This was followed by rock sampling. The floor was found to be underlaid by a longitudinal trough of relatively young (Triassic and Jurassic) rocks, related to those on the northern shore of the Channel and very different from the hard older rocks of the southern shore. The eastern portion of this trough extended into Somerset. The westernmost of its two synclines was followed until the rock floor was unreachable by the corer because of the thickness of the sand cover. The southern edge of the trough of young rocks was emphasised by an east-west thrust fault. The whole structure was displaced laterally by northwest to southeast trending tear faults, presumably of similar age to the larger ones dissecting southwest England. The markedly contrasting appearance of the old and young suites of rocks on sonographs helped to define their mutual boundary beyond where the rocks had been sampled. University workers dated the rocks and took additional rock samples while operating from other vessels.

For me, the most encouraging finding made with side-scan sonar during the early 1960s was that two sandy bedforms, of types that in the 1950s had been known only locally and from small areas, were now found to

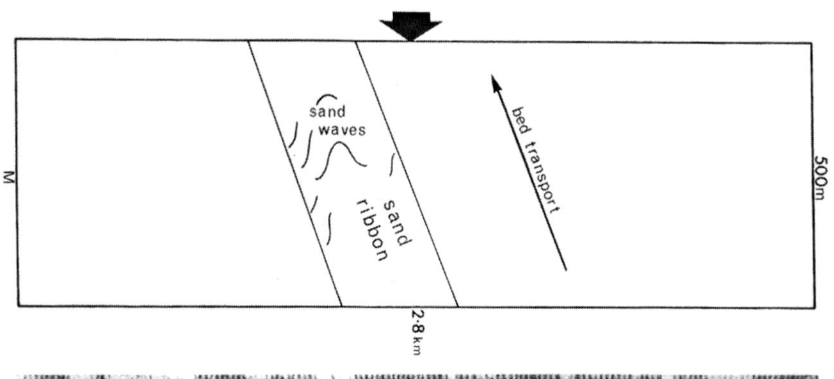

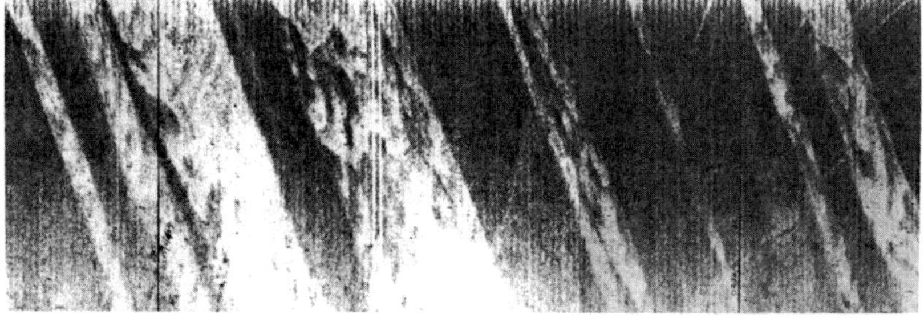

Sand ribbons (pale) overlying a gravel floor (dark) in the English Channel revealed by side-scan sonar (2.8 km by 500 m).

be widespread around the southern half of the British Isles. I hoped to understand the sedimentation processes occurring in this region, as well as other continental shelves with strong tidal currents. Recognition of these varied bed features and distinguishing between them and numerous extraneous patterns was difficult at first, but became easier as the amount of available information increased.

We started by studying sand waves, many of which were also shown on echosounder profiles. Trains of these, where one side of the crest was steeper than the other, were known to indicate their direction of advance in rivers, as sand swept off the gentle slope was deposited on the steeper slope ahead of it. In a sea with tidal currents, the relatively steeper slope indicated the net sand transport direction. The inferred migration direction of sand waves in the Irish Sea, the Bristol Channel and the English Channel corresponded with the direction of flow of the stronger ebb or flood tidal current. The difference in peak speeds between the ebb and flood tidal currents at any site was generally small, but significant.

The second widespread form of sand bed was the finger-like bodies up to 15 km long and 200 m wide that were laid out approximately parallel with peak tidal flows. The intervening ground could consist of gravel, as well as some stones, molluscan shells or shell debris. These sand ribbons, first discovered off East Anglia in 1953, are similar to the features laid out by sand-laden winds on stone-carpeted desert floors. The marine features help to define the paths followed by the sand, although they cannot indicate whether it is moving in the direction of the peak ebb or the peak flood tidal current at any location.

I used directional information from sand wave asymmetry along with peak tidal currents and sand ribbon elongation to map the directions of net sand transport around the southern half of the British Isles and neighbouring coasts of continental Europe. These paths were aligned more or less parallel with the coasts. But the direction of net sand transport was not the same throughout a given sea. For example, between North Wales and Ireland the net sand transport direction was northward in the north and southward in the south. Similarly, net sand transport was to the west in the western half of the English Channel, but to the east in the eastern half of the English Channel. Such diverging paths were also recognised in the Southern Bight of the North Sea between East Anglia and Holland. The grounds where these divergences took place were subsequently called bed-load partings. In Dover Strait, net sand transport directions were found to converge from the eastern part of the English Channel and from the North Sea, at what were subsequently called bed-load convergences, associated with a group of sand banks.

Net sand transport paths, associated with accompanying progressive decrease in grain size of the sediments in the direction of transport, revealed where the sand was being derived from and where it was now being deposited. These modern offshore tidal current sands were present on only some parts of the continental shelf around the southern half of the British Isles. There was a small area in the west-central part of the Irish Sea, as well as a narrow band to the south of that sea that extended eastwards towards the mouth of the Bristol Channel, also beyond the English Channel and west of Holland. These findings justified the view that discovering the location of modern deposits and distinguishing them from areas of non-deposition or erosion would require an understanding of the processes operating in the sea.[4]

The first detailed study of the sediments on the north-going net sand transport stream in the Irish Sea was conducted by Bob Belderson. Commercial short-range side-scan sonar became available and used from 1960 onwards. It was also used by other scientists at home and abroad including my one-time colleague Derek Chesterman, when he was examining the seafloor near Hong Kong.

By 1964, the short-range side-scan sonar had become a vital tool for our geological appraisal of the continental shelf. NIO sought funds to design and construct a long-range side-scan sonar to examine the deep seafloor. Much rough ground in the deep sea suitable for study by side-scan was known from echosounding, and a short-range device towed on the end of a very long cable by the Americans had already revealed the existence of interesting new features. However, the considerable drag on such a long cable meant that the towing speed could only be about 1 m/s, while regional surveys demanded much higher towing speeds. The design, construction and testing at sea of the new underwater vehicle – called GLORIA (Geological Long Range Inclined Asdic) was a major project, involving many NIO specialists. Survey results were not available until 1969.

Working Outwards

The wealth of short-range side-scan sonar data for the continental shelf now available encouraged a German archaeologist to ask us whether there were any drowned, ancient hut-circles on the seafloor of Mounts Bay, off Cornwall. It also enabled re-assessment of the bedforms on net sand transport paths, recognition of additional bedforms, as well as clearer ascertainment of areas of ongoing modern deposit accumulation. For example, Bob Belderson and I recognised a longitudinal sequence of bedforms by means of sonographs and sediment samples in the Bristol Channel:[5] this was subsequently found to apply in the other seas around the southern part of the British Isles. In the innermost zones of the Bristol Channel, where tidal currents are strongest, the clay layers of Jurassic rocks on the floor have been etched out, leaving the intervening more resistant stone-bands standing proud as ridges. Beyond this, going west, there were successive zones, first of coarse sediments such as gravel and stones with sand ribbons. Further beyond there were large sand waves, then a zone of small sand waves, followed by a flat sand floor with muddy sand beyond.

On the outlying ground, towards the Devon coast, modern deposits were represented by patches of sand resting on a floor of coarser materials. A detailed side-scan sonar and sampling survey nearer to the south coast of Ireland by Neil Kenyon found that there the sharply-defined sand patches were either transverse or longitudinal with respect to the peak tidal currents. However, the strength of the tidal currents was substantially less than found for such features elsewhere around the British Isles. It was tentatively concluded that the bed features must be formed and maintained when tidal currents are aided by the sizeable oscillatory bottom currents due to storm waves. Meanwhile, in 1968, Dutchman J.J.H.C. Houbolt removed any lingering doubts about the northward movement of sand waves past Holland. His cores showed conclusively that they had grown northwards by deposition of successive layers of sand on their relatively steep north-facing slopes.

In 1969, my colleague Mike Johnson estimated the amount of sand being transported past Holland in the offshore zone, taking account of the tidal and non-tidal currents and the oscillatory bottom currents due to wind-generated waves and swell. He concluded that when spring tides are aided by storm waves the sand transport will be ten times greater than during unaided spring tides. Indeed, more sand could be transported in a single day than during the preceding month. Thus, the offshore tidal current deposits in this region probably provide a record of extreme events rather than the lesser day-to-day events. Such deposition will continue as there is much sand still to be removed from west of Holland. Erosion of the East Anglian coast also provides new material for transport. In contrast, much of the floor of the English Channel has almost been stripped of its sand, although the molluscan fauna still provides shell debris.

The shape of the sand ribbons, seen on sonographs, varied with the strength of the maximum near-surface tidal currents.[6] By 1970, there was sufficient side-scan sonar coverage to enable Neil Kenyon and myself to provide a much more complete map of the net sand transport directions, extending the 1963 findings around the rest of the British Isles.[7] It showed the presence of additional bed-load partings and convergences. The net sand transport directions were supported by radioactive tracer studies in the North Sea. The overall pattern has now become the widely accepted model of modern offshore tidal current sedimentation. MBA workers R.D. Pingree and D.K. Griffiths subsequently showed that our map closely resembled their numerical simulation map of maximum bottom shear stress vectors occurring over the twice-daily tidal cycle.[8] This justified the empirical use of bedforms, despite limited information about the mechanisms generating them. Our map distinguishes sand ribbons, *senso stricto*, and sand patches. Sand ribbons are well formed, long and narrow, and are associated with stronger tidal currents than are needed to make sand waves. Sand patches are much shorter, ill-formed, longitudinal features, and associated with much weaker tidal currents. It is not yet clear whether these two longitudinal bed features are end members of a continuous sequence, or have different origins.

In 1971, we carried out a detailed survey of the bed-load parting region to the south of the Isle of Wight with the aim of exploring the possibility that the sand ribbons of strong tidal currents were associated with a new bedform, as suggested by available sonographs. This revealed the presence of narrow furrows up to 9 km long and up to about 1 m deep: their length aligned approximately parallel with peak tidal flows, with adjacent furrows joining in the direction of net sand transport. This direction of joining was to the west on the west side of the bed-load parting, but to the east on its eastern side. The furrows were therefore formed by the peak tidal currents in some way, rather than by rivers during a period of low Pleistocene sea level. A revealing finding was that some sand ribbons seem to originate at the distal ends of the longitudinal furrows, which must have acted as supply-route gutters. These furrows may be equivalent to the features seen on the eroding sand floor of one of the Edinburgh Channels that I noted in 1953, as well as to the furrows up to 4 km long and 1 m deep in the mud floor of Southampton Water, noted by my colleague Keith Dyer.

Sand patches were widespread on the continental shelf west of France, Ireland and Scotland, all regions of weak tidal currents. These bedforms were much smaller than could have been appreciated in the 1950s from navigational charts. In addition, on the outer part of the continental shelf west of Scotland[9] as well as around the edge of the Norwegian Trench, in the North Sea, our sonographs revealed a very different pattern of crisscrossing troughs gouged into coarse sediments. These troughs were up to 100 m wide and 2 m deep; one was followed for 5 km. They were interpreted as iceberg plough marks of the last glaciation of Pleistocene times. They are partly filled by modern sand or mud. Derek Chesterman and I showed that the iceberg plough marks off Denmark are being buried beneath sand brought in by the present-day northerly-directed coastal current.

In 1973, a broad review of sand and mud transport occurring in the North Sea refuted the belief of some workers that sand waves seaward of Holland were not present-day features. It also pointed to the likelihood of sand transport occurring beyond the sand wave zone on occasions when large storm waves were able to provide significantly strong oscillatory bottom currents to aid the tidal currents. It contrasted the different shapes of the sand accumulations occurring at three bed-load convergences and suggested that their form may depend on the speed of the peak tidal currents.[10]

The continental shelf of north-west Europe is an excellent region in which to investigate how present-day water movements produce modern offshore tidal current deposits. The progressive recognition of a suite of bedforms indicating sand transport paths and the directions

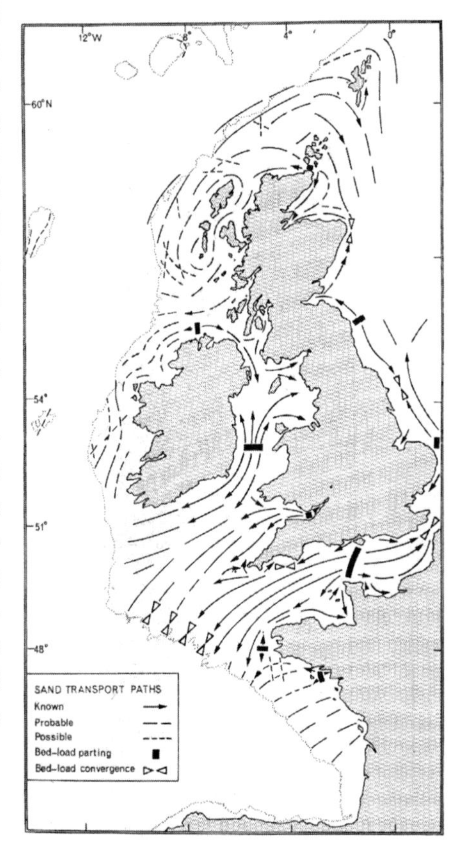

Sand transport patterns on the European continental shelf derived from side-scan sonar records.[7]

of net sand transport, revealed both the sources of the sand and the location of the modern deposits. The internal structures of these deposits need to be determined before they can be used effectively to interpret analogous deposits laid down in shallow seas of the geological past. Present work makes a vital distinction between sediments that have reached their final resting place as deposits under existing conditions, and material that is still in transit. Of course, the present sedimentation pattern is but a stage in the evolution of these seas. Changes in basin shape and dimensions will affect the tidal currents and be reflected in the nature and location of future deposits.

There has been relatively little deposition of sediment on the continental slope and adjacent continental shelf between Scotland and Portugal, since the North Atlantic was formed by seafloor spreading. This was shown by continuous reflection profiles obtained on a cruise of *Discovery* during 1965, with Joe Curray of Scripps, Dave Moore of the Marine Environment Division of the US Naval Underwater Weapons Center (NUWC), Bob Belderson and myself.[11] The profiles also showed the massive amount of erosion that had taken place during periods of low Pleistocene sea levels when the European rivers, flowing over the continental shelf, discharged their sediment loads close to its edge. Turbidity currents cut the numerous submarine canyons. Faults and slumps helped with the destruction which, in places, has cut down more than 1 km through Cretaceous and Tertiary strata to expose pre-Cretaceous rocks of basement-aspect at the foot of the continental slope. The profiles also provide a glimpse of an earlier phase of erosion, somewhere between the Cretaceous and Tertiary phases of deposition.

A traverse with GLORIA along the edge of the continental shelf west of Brittany with the side-scan sonar beam pointing down the slope provided a 'plan' view of the relief of submarine canyons. The axes of the canyons were obvious. The sonograph showed the presence of numerous gulleys in the canyon walls, many of which commenced on the un-dissected portion of the continental slope between the canyons. This implies that sediment was entering both the heads of the canyons and their sides. In some places, portions of the canyon walls had slid away as slumps and, where these were impeded, slump folds seem to have developed.

South of Portugal, sonographs of the relatively shallow ground of the Gulf of Cadiz enabled Kenyon and Belderson to provide information about the undercurrent that flows out from the Mediterranean through Gibraltar Strait.[12] The sonographs revealed bedforms in the Gulf of Cadiz, already familiar from the tidal current-swept seas around the British Isles, included longitudinal furrows in the Gibraltar Strait, where the current was strongest. To the west, familiar sand ribbons were found; further beyond there were sand waves, and finally there were mud waves up to

40 m high (the only bedform not known from around the British Isles). These bedforms implied the same westerly direction of flow of the near-bottom current as measured by previous workers using current meters. The sonographs also provided more extensive coverage, including the limits of the peak flow, as well as showing its progressive westerly decrease in peak strength as the current spread out. In addition, where the flow was constrained by rock ridges (outlined using GLORIA) the presence of sand waves showed that there had been a local increase in current strength. Overall, the side-scan sonar provided a quick and easy way of defining the approximate limits of the main flow and could have guided physical oceanographers as to where best to locate their current meters. However, such an approach is only worthwhile where sand is being fed into the current system. In the present case, the near-surface current flowing into the Mediterranean, aided by the tidal currents on the continental shelf, is moving sand towards the strait and dumping it in the deep water where it is swept away westwards. Based on this, we advised a company responsible for an underwater cable across the Gulf of Cadiz that it would be advantageous to move it away from one of these narrow channels where it was being damaged repeatedly. In the Eastern Mediterranean, the first sonograph of the so-called cobblestone topography showed that it actually consists of linear and curvilinear structures that result from compression of sediments that have been caught up between the northward-moving African Plate and the European Plate. Subsequently these structures were the subject of extensive study.

A book entitled *Sonographs of the Sea Floor* was published in 1972 to alert researchers to the value of side-scan sonar for examining the seafloor.[13] It is primarily concerned with recognition of geological features but also includes such man-made items as wrecks, pipelines and rigs, together with marine life and sea effects. The 163 sonographs illustrated single features, each with accompanying interpretive diagram. The book includes some of the earliest sonographs obtained by GLORIA, showing submarine channels on volcanoes, cliffs and ridges.

15

The rocks below the deep ocean

Anthony Laughton

What we knew

When NIO was created in 1949, it had the mandate to study "*the science of oceanography in all its aspects*". Dr Deacon expanded the research base at NIO from waves, tides and currents, whales and marine biology, to include both marine geology and marine chemistry. As well as a programme of research on the continental shelves, Deacon wanted the work of NIO to include the deep ocean floor.

I was recruited to NIO in 1955, after I had spent a year at the Lamont Geological Observatory in the USA. Deacon asked me at a meeting in Woods Hole what I was doing after I returned to the UK, and said that if I failed to get a job in Cambridge I should ring him at NIO. So I did, and he said "*Come next Monday*". When I arrived he asked me what I wanted to do, and I replied that the UK should investigate the ocean floor with photography, to which he replied "*Go ahead*". This was Deacon's style of recruitment in those balmy days.

He was well aware of the pioneering work being done in marine geophysics at the Department of Geodesy and Geophysics (G&G) at Cambridge University under the leadership of Sir Edward Bullard, who had served on the original Oceanography Sub-Committee, and of Dr Maurice Hill. He also was in touch, through his wide international connections, with the work being carried out by the Lamont Geological Observatory of Columbia University in New York led by Prof. Maurice Ewing and with the work of Dr Bill Menard (formerly of the Naval Electronics Laboratory) at the Scripps Institution of Oceanography under the Director, Prof. Roger Revelle. Deep-ocean geology and geophysics had benefited enormously from the peacetime exploitation of techniques developed for submarine warfare. The extensive use of sonar, the refinement of echosounding, the measurement of magnetic and gravity fields at sea, and the use of underwater explosions were all put to use in determining the shape and composition of the deep-sea floor and the structure beneath it.

At Cambridge, Hill took over from the earlier pre-war marine seismic refraction techniques that Bullard had used with Tom Gaskell, and which had been pioneered by Ewing. In contrast to the American system of using two ships for refraction seismics, one to shoot and the other to receive, Hill developed a one-ship technique using sonoradio buoys which were laid out in a line to receive signals from the shots and radio them back to the shooting ship. This system was much more economic and flexible in regard to ship time.

While a research student at G&G, under the supervision of Hill, I took part in cruises using this seismic refraction technique, as well as collecting sediment for my thesis on the changes in the velocity of seismic waves in compacting ocean sediment. Hill had interpreted his seismic data as showing that there was a velocity gradient in the sediments. I was able to reproduce this in laboratory experiments.

It is worth considering what was known of the deep ocean floor at the time of the early days of NIO, and of the ideas about how it was created. The very early theories about continental drift of Wegener, and later of du Toit, Arthur Holmes and Warren Carey, in which it was envisaged that the continents had drifted apart floating in a 'sea' of magma, were discredited by Prof. Harold Jeffreys, the *eminence grise* in planetary geophysics of the early 1950s, who had firmly demonstrated from global seismology, that the mantle was too strong to enable Wegener-type continental drift to occur. Jeffreys' book, *The Earth,* was obligatory reading for any student starting at that time. Geologists believed that continents could move up and down but not sideways.

The 'Bible' for oceanographers was the book, *The Oceans – their physics, chemistry and general biology* by Sverdrup, Johnson and Fleming first published in 1942, and reissued several times during the following twelve years. Indeed Deacon's Secretary, Miss Thomas, used the chapter headings of the book as the basis for her filing system. Although there are chapters concerned with the topography of the ocean floors and with marine sedimentation, there is little about the structure or origins of the ocean basins.

The work of Lamont, Scripps and Cambridge, collecting sounding and geophysical data, and samples of the deep seabed, began to show that below the sediments it was predominantly made of basalt and that the ocean crust was very much thinner than continental crust. In the late 1950s Bruce Heezen, Marie Tharp and Maurice Ewing at Lamont demonstrated that there was a mid-oceanic ridge encircling the world, and that associated with it was a valley along its median line. Along the axis of the median valley was a concentration of shallow earthquakes. Their classic book, *The Floor of the Oceans – 1. the North Atlantic,*[1] published by the Geological Society of America in 1959, summarized what was

Onboard Discovery II *in 1956. (Left to right)* ***Back row:*** *John Swallow, Maurice Hill, John Cleverley.* ***Front row:*** *Anthony Laughton, Capt Joe Dalgleish, Sir Edward Bullard.*

then known about the seafloor. The final paragraph of this book, headed '*The origin of the Mid-Atlantic Ridge*', concludes "*we are still a very long way from having a comprehensive knowledge of the Ridge*" and that "*the topography of the Mid-Atlantic Ridge is largely the result of normal faulting. Whether the forces are the result of horizontal extension or vertical uplift remains the most important unsolved problem in connection with the origin of the continental as well as the sub-oceanic rift valley systems*". The hypothesis of seafloor spreading and the now well-accepted concepts of plate tectonics were just around the corner.

Deacon, recognising the need for the UK to contribute to the exploration of the deep ocean floor, made *Discovery II* available to the G&G at Cambridge for a month's cruise every other year, an arrangement that continued for several years with NIO sharing cruises with G&G.

At about this time, John Swallow, who was then another research student at G&G and who also was later to join NIO, was circling the world on HMS *Challenger* with Tom Gaskell, and initially Hill, to make seismic refraction measurements in the Atlantic, Pacific and Indian Oceans. Together with data from Prof. Russ Raitt and George Shor in Scripps, a picture was emerging of an oceanic structure that was significantly different from that of the continents.

Underwater photography

At Lamont, I had been involved closely with its newly-designed deep-sea camera, and so, with the engineering help of Dicky Dobson and later John Jopling and Peter Collins, I designed and built at NIO an acoustically monitored and pressurised camera and flash.[2] This became the basis for underwater photography used not only by me, but also by the biologists and the physical oceanographers. It may seem strange now that one could not have bought a camera off the shelf, but in the 1950s, there was little marine equipment on the market and anything which was needed for oceanographic research had to be designed and built from scratch. The Applied Physics, and Instrument and Engineering, groups at NIO were crucial to achieving this. From my experience in the USA, I was able to introduce the concept of O-rings to NIO for sealing against high pressure and to persuade the engineers of their effectiveness.

Eventually deep-sea cameras were made at NIO for other laboratories around the world but it was clear that NIO did not have the after-sales service needed to maintain this commercial approach.

Deep-sea photographs showed that the seabed, far from being the passive place where there was no life and where the currents were very weak, sustained a live benthos and a population of near-bottom fish, was covered with mounds and burrows, and in places was marked by ripples in sand indicating active bottom currents. Exposed rocks were seen on seamounts and in particular on the Mid-Atlantic Ridge which showed fresh glassy basalts and little sediment. Occasionally there

Tony Laughton with the deep-sea underwater camera, showing the pinger (top), the camera (middle) and the flashlight (bottom).

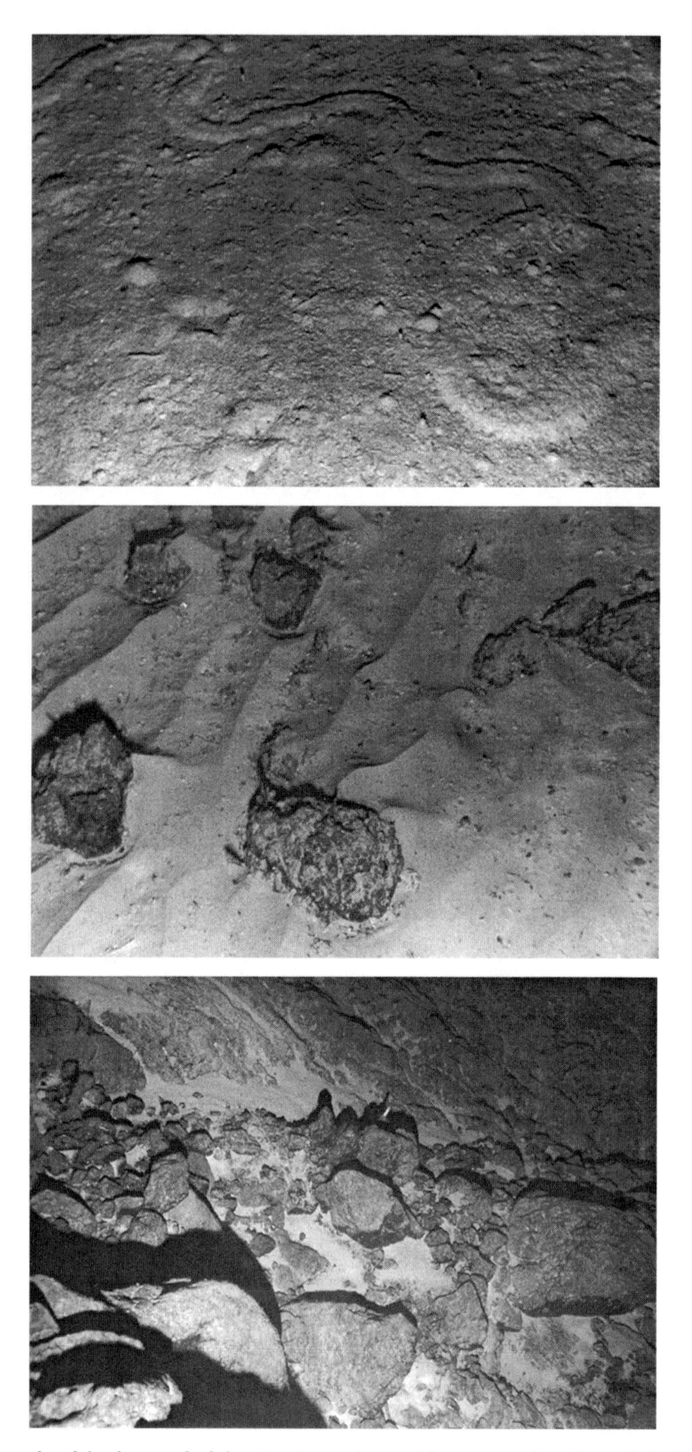

Photographs of the deep sea bed showing (top to bottom) burrows and tracks made by bottom living creatures at 5,282 m, sand ripples indicating strong currents at about 3,000 m on a seamount and rocky slopes at the bottom of Palmer Ridge at about 3,400 m.

were fields of manganese nodules covering the seabed. A book, *Deep-Sea Photography*[3] was edited by Brackett Hersey of WHOI in 1967, bringing together photographs of the seafloor from all over the world, including NIO.

Deep-sea corer being lowered over the side.

In 1956 and in the following ten years, I collaborated closely with the Cambridge group under Hill taking part in seven cruises in *Discovery II* and later in its replacement RRS *Discovery*. These cruises examined various features of the North Atlantic and northwest Indian Ocean, mapping them, sampling by coring and dredging, measuring magnetic and gravity fields and deep structure with seismic methods, all with a view to understanding the origin of the local and regional geology. We were able to map in detail a part of the median valley of the Mid-Atlantic Ridge, which Swallow had surveyed in 1949 during the *Challenger* cruise, and photograph fresh basaltic rocks on its flanks.

We sampled the rocks of the partly buried seamounts in the ocean basins, showing them to be mainly basalt but scattered also with glacial erratics. Abyssal plains were mapped and sediment cores showed how they were formed by turbidity currents, sometimes very violent, arising from 'landslides' on the continental slopes and carrying sand, silt and mud from the continental margins, progressively filling the ocean basins.[4,5] From cores in the Madeira Abyssal Plain, Belderson and I were able to correlate individual turbidite layers over a distance of 65 km.[6] When one basin was filled it overflowed into the next, eroding an interplain channel between the two. The eroded sides of the channel allowed the structure of the abyssal plain to be examined.[7] Steadily the geological processes on the deep-ocean floor were being revealed.

Charting the ocean floor

Working with the Cambridge group at sea, my role was to take photographs of the seabed and, as an essential support for the geological and geophysical work, to prepare contour charts of the topography (the shape of the seafloor), which were effectively proxy maps of the geology. Deep-

Rock dredge. The bag is made of chain links.

sea soundings were at that time collected mainly by research ships run by academia and by hydrographic ships run by navies, the only ships that were equipped with deep-sea echosounders. The soundings in the North Atlantic were compiled by the Hydrographic Department in the UK on 1:1 million-scale sheets, and it was these that I undertook to contour, together with David Roberts and cartographers in NIO, bringing to bear the increasing knowledge of the processes on the seabed. These sheets, covering much of the NE Atlantic Ocean, were compiled and published as coloured charts,[8] and have subsequently provided the basis for much of the oceanographic and geological research in the area.

In the 1950s, the deep-sea echosounders were of poor quality and inaccurate. Swallow and I, using American experience, built in 1958 a precision echosounder at NIO, based on the linear and precisely scanning Mufax recorder. This much-improved equipment enabled the abyssal plains to be contoured in places at one fathom intervals.

Internationally, soundings from all over the world were being contoured in a programme, the General Bathymetric Chart of the Oceans (GEBCO), which had started under the auspices of Prince Albert 1st of Monaco in 1903.[9] At NIO, Henry Herdman was a member of the GEBCO Committee, and I took over his role in 1966 and enabled NIO to continue to play a significant role in global charting. A paper in 1973 by Laughton, Roberts and Graves,[10] and a subsequent meeting at NIO, had a significant impact on the future of mapping within GEBCO, by specifying precisely how bathymetry (the measurement of depth) should be portrayed and how GEBCO should be managed. The new

A successful haul of rocks from the dredge.

management structure involved the Intergovernmental Oceanographic Commission, representing scientists, to work together with the International Hydrographic Organisation, representing navigation chart makers. GEBCO continues today to collect and contour deep-sea soundings from all over the world, now using digital techniques.

Tectonics of the ocean crust

Another summary of the state of knowledge of oceanography, *The Sea*, was published initially in three volumes between 1962 and 1963, edited by Hill.[11] Although there was much on the geology of the sea floor, the authors of the chapters on topography, on the magnetic field and on the mid-ocean ridge still were struggling to find a mechanism for the evolution of the ocean basins. Heezen and Ewing stated that "*most current workers believe that the mid-ocean ridge is the result of tension and that material rising from the mantle beneath the crest is adding new rock to the floor of the rift valley*". Robert Dietz in the USA, building on earlier ideas from Harry Hess, had published his theory of 'the spreading of the seafloor' in 1961 which was being confirmed by subsequent geological surveys of the seabed.

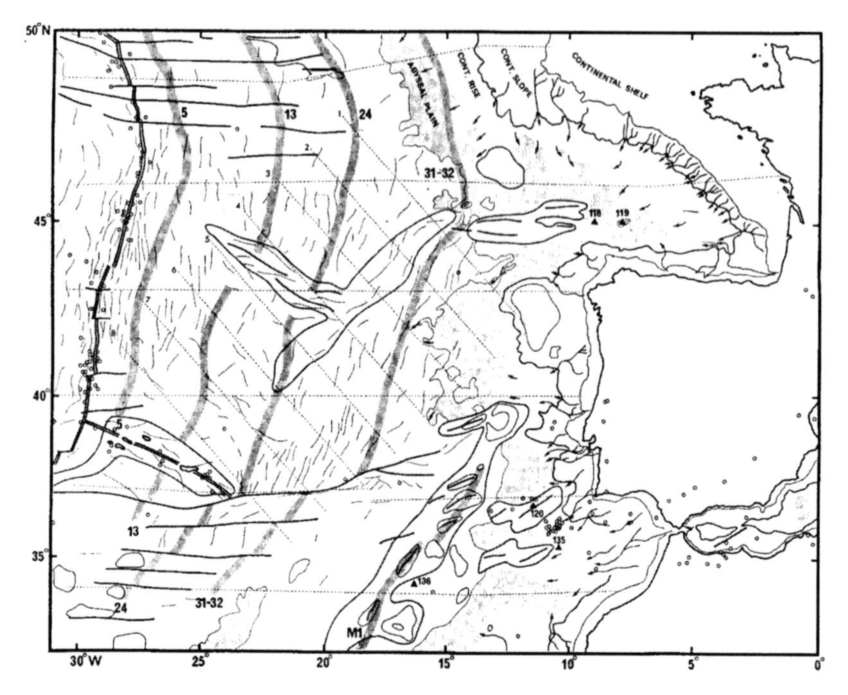

Physiographic features of the NE Atlantic showing the continental shelf and slope, faults and fracture zones. The principal magnetic anomalies, (numbered), parallel to the median valley, are shown as shaded lines.

By the time Volume 4 of *The Sea* was published in 1971,[12] new concepts of the evolution of the seafloor had revolutionised not only marine geology, but also continental geology. In 1963 Fred Vine and Drummond Matthews at G&G had related seafloor spreading to the random reversals of the earth's magnetic field to explain the linear magnetic anomaly patterns mapped at the sea surface, enabling the rate of seafloor spreading to be determined from the spacing of the stripes. The symmetry of the pattern of stripes either side of spreading centres in all the oceans gave strong confirmation of the universal role of seafloor spreading. Together with the understanding of subduction zones, this formed the basis for the plate tectonics hypothesis.

It was against this background of exciting and groundbreaking new understanding of the geological evolution of ocean basins that the research programme at NIO was conducted in collaboration with Cambridge, and later on its own. Many of the geological and geophysical cruises collected data that helped support and consolidate the ideas of plate tectonics and ocean basin evolution. Although the results of individual research stand out as milestones in its development, it was the collective knowledge of marine geologists and geophysicists throughout the world that gradually confirmed the status of plate tectonics.

The International Indian Ocean Expedition (IIOE)

It was partly to test the concept of seafloor spreading that I and others at NIO and at Cambridge organised a cruise to the Indian Ocean as part of the IIOE. Deacon had said to me in 1958, "*What do you know about the Indian Ocean?*" to which I replied "*Rather little*". "*Can you think up a plan for the UK to undertake on geological problems there as part of an international programme*? *Don't work too hard*" he said as he left the office.

The IIOE was the first major international collaborative effort in oceanography following the success of the International Geophysical Year, and was generated by SCOR (at that time the Special Committee for Oceanic Research), of which Deacon was Vice Chairman. I was closely involved in the planning of both the national and international aspects of the geological research.

It was clear from the work of Heezen that a mid-ocean ridge ran south from the mid-Atlantic around South Africa and linked with the mid-Indian Ridge and the Carlsberg Ridge in the Arabian Sea. If seafloor spreading was a universal process the westward extension of the Carlsberg Ridge into the Gulf of Aden and the Red Sea should show how the process could split apart continents and generate new oceans.

The geological and geophysical work in 1963 in *Discovery*, led by Hill, was concentrated on the NW Indian Ocean with special emphasis on the Carlsberg Ridge and its western extension. Preliminary work by HMS *Owen* making long profiles across the Carlsberg Ridge had shown a seamount, which Fred Vine from Cambridge later surveyed in detail, which turned out to be magnetised in a reverse direction relative to the Earth's present field. It was following this discovery that Vine and Matthews, in 1963, published their seminal paper.

I joined Hill and some of his research students and others from NIO, for the 1963 cruise to the NW Indian Ocean. This led me to my first research on the origin of the Gulf of Aden, using data from passage tracks to map the morphology typical of a mid-ocean ridge and to identify a central magnetic anomaly. I was able to demonstrate the geological linkages on the continents either side of the Gulf and show the direction of relative movement as the Gulf split apart, starting in the Miocene. [13] The topographic charts showed oblique transform faults crossing the Gulf and how, at its eastern end, the central ridge was offset to the south by the newly discovered Owen Fracture Zone, to carry on in a southeasterly direction as the Carlsberg Ridge.

At its western end the split penetrates into the Afar Depression through the Gulf of Tadjura. Prior to separation of Arabia from the Horn of Africa the region arched up and subsequently dropped by block faulting. The origin of the Gulf was linked to that of the Red Sea. In an article in the *New Scientist* in 1966, I was able to show that Arabia had rotated away from Africa to create the Gulf and the Red Sea.

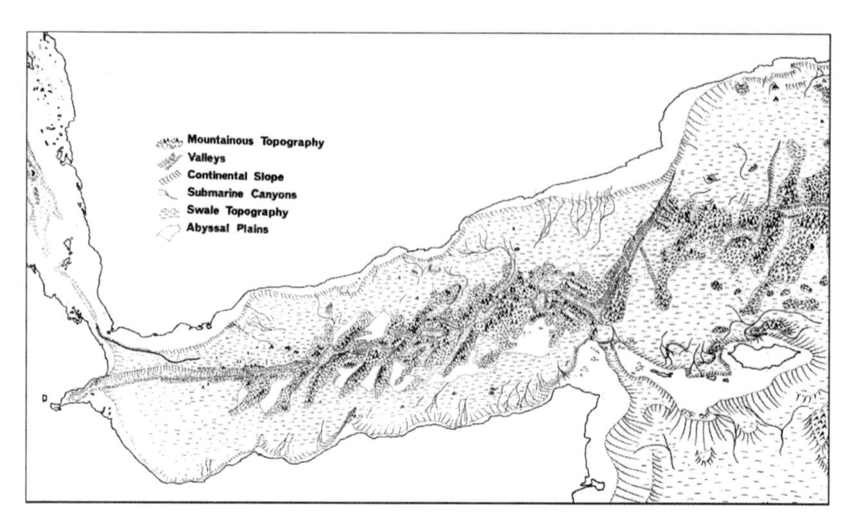

Diagram showing the shape of the seabed in the Gulf of Aden.

These results were later presented in December 1964, at the International Geological Congress (IGC) in New Delhi. As a result of pressure from the British National Committee for Geology (a Royal Society committee, of which I was a member), the IGC had decided to include a session on marine geology which had hitherto not been one of their sessional topics! As both Matthews and I were planning to attend the Congress we had each prepared three papers for presentation on results from the *Discovery* cruise in the Indian Ocean. One of these was on my Gulf of Aden work; others were on the crustal structure of the ocean floor determined by the seismic refraction section between Kenya and the Seychelles, and on underwater photographs of the Carlsberg Ridge. Unfortunately, while I was on a geological field trip to Kashmir, a violent snowstorm isolated the Kashmir party for the duration of the conference in Delhi, and while Matthews gave some of his papers in Delhi, I did what I could to a captive audience, some eighty scientists, in presenting the first results of the IIOE marine geology. I was able to spend an hour and a half talking about the origin of the Gulf, whereas in Delhi I would have had only twenty minutes!

A later NIO *Discovery* cruise in 1967 to the Gulf of Aden filled in much of the detail of the geophysical nature of the Gulf. The interpretation of the new results were presented at a two-day Discussion Meeting at the Royal Society in March 1969 and were later published in October 1970 in a paper entitled *The Evolution of the Gulf of Aden.*[14] I was able, with Bob Whitmarsh and Meirion Jones, both of NIO, to give a much more detailed analysis of the phases of the growth of the Gulf of Aden, dating the magnetic patterns and mapping the topographic trends, as well as publishing new bathymetric and magnetic charts of the Gulf in colour. In the same publication I also presented a new bathymetric chart of the

Red Sea, which I had prepared for the many geophysical and geological studies which were then being made of the region.

One of the outcomes of the IIOE was the preparation and publication of an atlas summarising the achievements in marine geology and geophysics. Thirteen countries and forty-six research ships had taken part over the six-year period of the IIOE between 1959 and 1965 and had collected an enormous amount of data. I was one of six editors for the *Geological-Geophysical Atlas of the Indian Ocean* together with Robert (Bob) Fisher of Scripps, Eric Simpson of South Africa, Victor Kanaev and Dina Zhiv of the USSR, working under the Chief Editor, Gleb Udintsev, also of the USSR. The atlas was finally published in Moscow in English and in Russian in 1975, comprising over 150 pages in an extremely large format to accommodate the charts.[15]

GLORIA, a deep-sea side-scan sonar

Following the very successful use by Stride of the hull-mounted side-scan sonar on *Discovery II* to determine the morphology of the continental shelf out to a range of about 1,000 m, Deacon was persuaded in 1965 that the side-scan sonar principle could be scaled up to give similar geo-morphological information about the deep seabed. Work started on the Geological Long Range Inclined ASDIC, or GLORIA as it immediately became known. Details of this major construction are given in the following chapter. The seabed could be surveyed at cruising speed out to 20 km to the side of the ship.

During the period up to 1975, I led a number of *Discovery* cruises exploiting this new technology. On Cruise 43 in 1971, we surveyed the plate boundary between the Eurasian and African tectonic plates, that runs between the Azores and SW Spain. It turned out that near the Azores this

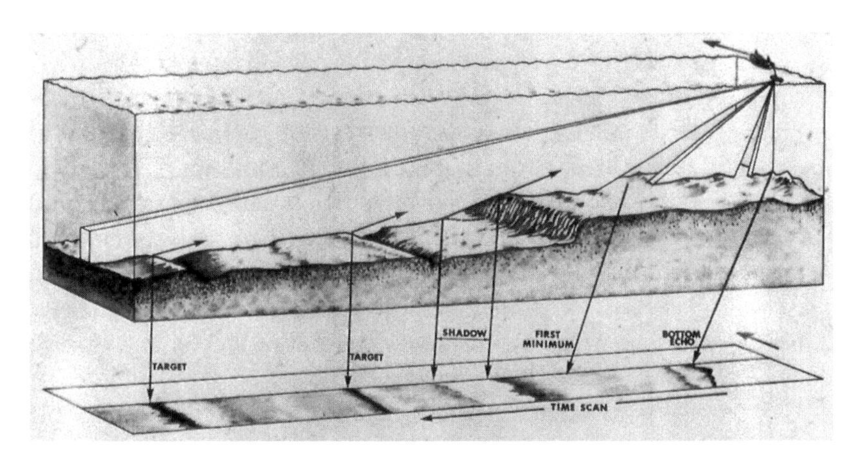

Diagram of method of scanning the seabed by GLORIA

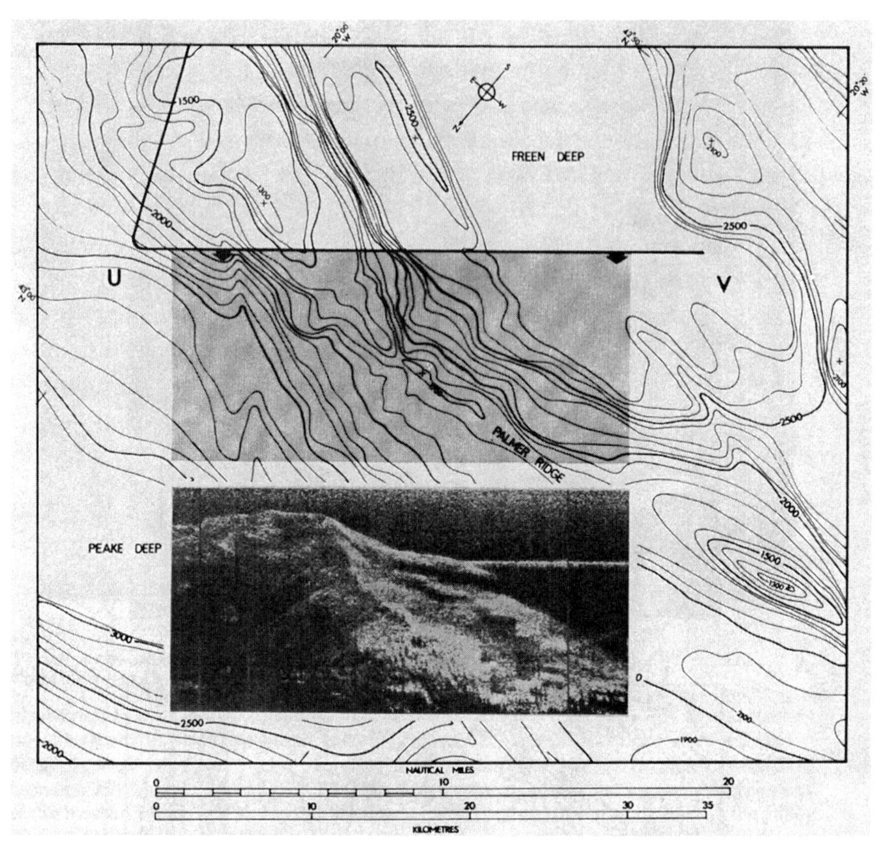

GLORIA scan of Palmer Ridge (43°N, 20°W) compared with the bathymetry mapped independently. Palmer Ridge stands 2,800 m above the floor of Freen Deep.

was a long very narrow single deep fault, which we named the GLORIA fault, where the plates slid past one another.[16] On the same cruise we examined once more the twin troughs of Peake and Freen Deeps in the NE Atlantic, demonstrating dramatically how GLORIA could give an acoustic image in one sweep of the entire Palmer Ridge which separates them.[17]

In 1973, NIO contributed on *Discovery* Cruise 54 to the Franco-American Mid-Ocean Undersea Study (FAMOUS) project on the axis of the Mid-Atlantic Ridge southwest of the Azores. This region, straddling the median valley and including two offsetting fracture zones, had been surveyed in great detail by the US Navy using their multibeam swath bathymetry echosounder system. The results of this saturation survey, previously classified, were released for scientific use in this joint French-American exercise where submersibles were used for the first time to explore the details of an active spreading centre.

Overlapping swaths of GLORIA images, viewed always to the northwest, built up a mosaic of some 30,000 km^2 in the FAMOUS area which clearly showed the tectonic fabric of this slow-spreading centre.

The GLORIA towed vehicle ready to be brought back on board. Note the divers assisting.

Either side of the median valley, which contained a line of young and active volcanoes, the ocean crust was faulted into numerous parallel tilted blocks, all with their steep scarps facing inwards to the median valley. The faults were spaced by only a few km, but extended tens of km parallel to the valley, giving a striking texture to the ridge, which was interpreted in terms of the rising mantle under the ridge axis. The GLORIA images were calibrated by the very detailed survey done by the US and also related to the submersible dives and to underwater photographs that we took in the valley.[18,19] The study of the structure of mid-ocean ridges later became the speciality of Roger Searle who had by now joined my group.

Many of the sonographs obtained by NIO in the first decade with GLORIA are published in a review, *The First Decade of GLORIA.*[20]

GLORIA Mk I had provided unique views of the morphology of the sea bottom which complemented the contoured charts of echosounding and which enabled very large areas of the seabed to be surveyed very rapidly. The use of oblique insonification enabled features to be seen and interpreted that could not be detected on echosounding profiles or surveys. Highlights and shadows in the images produced pictures similar to those seen from an aeroplane in a setting sun. Even swath mapping using multibeam echosounders did not show all the features, such as meandering channels (presaging their recognition from 3D seismic surveys by fifteen or more years), that could clearly be seen with GLORIA. No other country had developed this technique and NIO led the way internationally for later generations of side-scan sonars.

It is hard to overestimate the role that side-scan sonar has had subsequently, not only in marine geology but also in marine engineering,

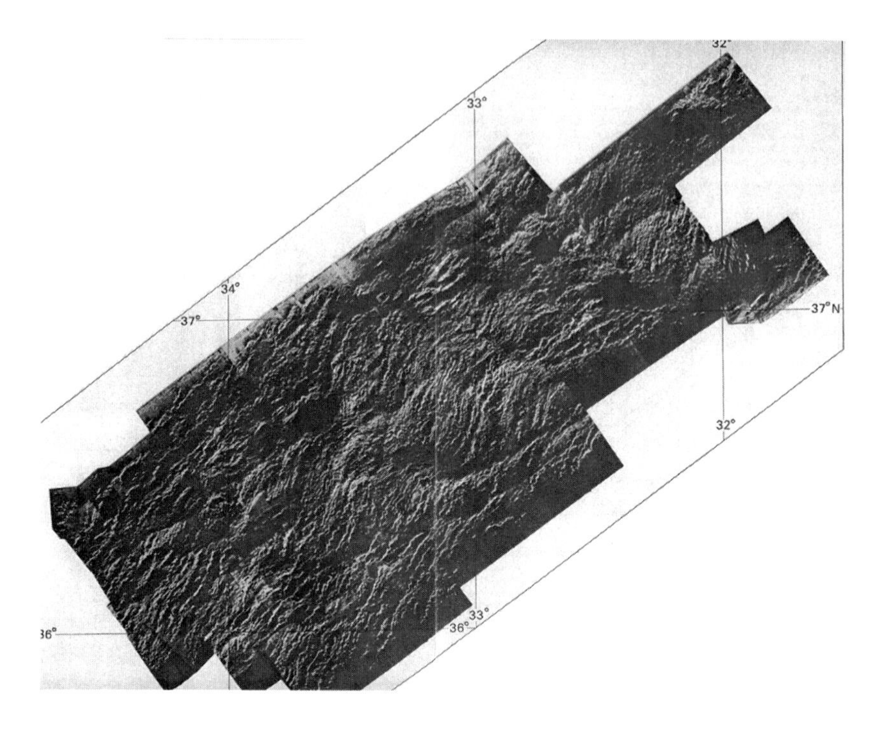

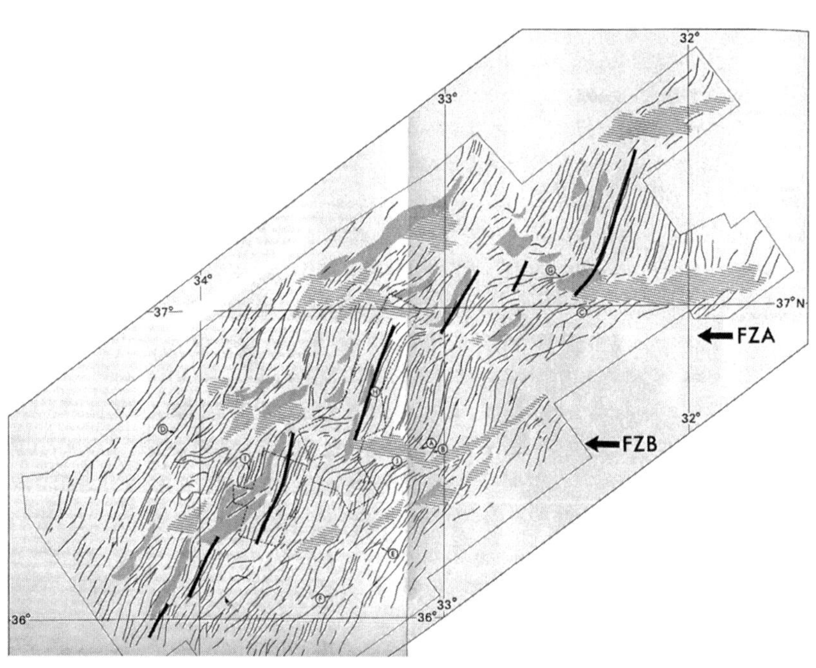

Top: *Mosaic of overlapping GLORIA images across the axis of the Mid-Atlantic Ridge.*
Bottom: *Interpretation of linear faults paralleling the median valley (heavy line) which is offset by fracture zones.*

wreck identification, site surveys for production platforms and commercial exploitation of the seabed. Industry eventually took over the design, manufacture and marketing of side-scan sonar systems.

The Deep Sea Drilling Project

In 1957 a group of distinguished geologists and geophysicists were having a wine breakfast in Scripps at the house of Walter Munk, meeting as members of the rather curious and flippantly named AMSOC – the American Miscellaneous Society. They were discussing an outrageous proposal by Munk at an earlier meeting of the National Science Foundation, where members were bemoaning the lack of new visionary grant applications, to sample the mantle by drilling a hole down to it. Prof. Harry Hess thought that this idea should be discussed by AMSOC which then pushed through a resolution, with the help of Dr Tom Gaskell of the UK, for the International Union of Geodesy and Geophysics to initiate a feasibility study.

So was born a vast project, the Mohole project (the name derived from the attempt to drill a hole to the Mohorovičič discontinuity, the boundary between the crust and the mantle), that was to consume huge sums of research funds and to inflate in cost until it was killed by the US Congress. But it was to spawn a lasting but less ambitious programme of drilling holes into the ocean floor. I was to find myself deeply involved in this latter programme, the Deep Sea Drilling Project (DSDP), in 1969.

While the fate of Mohole was being argued about, there was a growing body of scientific opinion that a single deep drill hole at enormous expense was not the right way to proceed. Several leaders of US oceanographic laboratories collectively proposed a programme of drilling many holes into the sediments of the ocean floor, and later possibly into the volcanic rocks that lay beneath.

In 1965, five US research institutions formed themselves into a group with the common aim of achieving this, under the acronym of JOIDES – the Joint Oceanographic Institutions Deep Earth Sampling group. Initially they contracted a drilling barge, the *Caldrill*, which anchored on the relatively shallow Blake Plateau off the east coast of the USA, to drill some shallow holes into the sediments. The samples were enthusiastically examined by the scientists.

Spurred on by the success of this, JOIDES set up the DSDP awarding a contract to Scripps to manage both the science and the drilling vessel, DV *Glomar Challenger*, which was contracted from Global Marine Development Inc. The technique of dynamic positioning, by the use of bow and stern thrusters and the main engines, relative to acoustic beacons on the sea floor had by now been refined and was integral to the operations of DSDP.

On the first leg, in 1968, holes were drilled on the Sigsbee Knolls rising out of the flat floor of the Gulf of Mexico, obtaining good sediment cores of salt, but also potentially dangerous shows of hydrocarbon. This success allowed subsequent planning to go ahead and thereby initiated the first of a series of ocean drilling programmes that continues today as the Integrated Ocean Drilling Program.

By 1969, JOIDES had started to discuss a two-month leg to look at geological problems in the N Atlantic. NIO had been watching the results and achievements of previous legs with great interest and felt that if the programme was to bring the drilling capability into the N Atlantic, we should make a bid to be involved. The Director of Scripps, Bill Nierenberg, was visiting Cambridge at this time and strongly encouraged the UK to take part.

The Atlantic Panel of the JOIDES Planning Committee had been looking at possible sites in the N Atlantic, and examined bids by various marine research groups, hitherto all American, to drill holes to answer well-formulated scientific questions. Not only had the questions to be significant, but there had to be adequate survey data to support the proposals and to define precisely where the holes should be drilled.

In the UK, NIO had a number of projects on the boil on the geology of the NE Atlantic with a considerable quantity of data to back them up, and much additional new data from European institutions was shown at a SCOR meeting on Atlantic margins in early 1970. In October 1969, I wrote to Mel Peterson, the Chief Scientist of DSDP at Scripps, suggesting that the UK could contribute to the N Atlantic component of the programme and was encouraged by him to formulate a UK proposal.

In November and December 1969, I organised two meetings of UK marine geologists and geophysicists at the Royal Society to put together proposals for nine possible sites, supported by data acquired over the last few years. The JOIDES Atlantic Advisory Panel, chaired by Ewing of Lamont, was enthusiastic about these and invited me to present them to the Panel meeting at Lamont in January 1970. In a somewhat chaotic meeting, at which possible sites were thrown randomly into the ring with little backing data, I argued for the UK proposals, five of which were accepted for drilling on DSDP Leg 12 later that summer. With the prospect of participating in DSDP, in April and May 1970, NIO made several site surveys of potential drill sites on the Rockall Plateau.

I was asked to be one of the Co-Chief Scientists on board, and was joined by Bob Whitmarsh from NIO. The other Co-Chief Scientist onboard was Bill Berggren, a micropalaeontologist from WHOI, whose skills in palaeontology and sedimentology complemented mine in geophysics.

Present configuration of N Atlantic showing the older magnetic anomalies (as heavy lines), and drill sites of DSDP Leg 12.

The team of ten scientists, supported by twelve technicians, comprised a mixture of disciplines and nationalities, and the cruise was an eye-opening educational experience.

Leg 12 in 1970 took us northward and eastward around the N Atlantic from Boston ending up at Lisbon. The first scientific target was an isolated and relatively shallow area north of the Grand Banks called Orphan Knoll. Was it a volcano or part of the breakup of Pangea as the Atlantic Ocean was created? Our cores showed it to be a sunken fragment of continent, left behind as the great crack between the combined European and Greenland land masses separated them from the American land mass.

Our scientific objectives continued with drilling in the Labrador basin to establish the age and nature of the initial split between Greenland and Labrador, next on to the flanks of the Reykjanes Ridge and to the Hatton-Rockall Basin on Rockall Plateau between Hatton and Rockall Banks, thought to be a piece of sunken continent, and then into the deep basin of the Bay of Biscay to try to date its origin. All of these holes were predominantly to refine the ideas on the evolution of the north Atlantic through its various stages. The data that we acquired was to provide the basis for many publications by all the scientific staff and by the wider community.

The conditions of participating in the DSDP required us to write up shipboard reports on the core material before we left the ship at our final destination, followed within the next year or two with extensive *Initial Reports of the DSDP*. I was able to bring together much of the drilling, topographic and magnetic survey data for the North Atlantic and to publish a paper entitled *South Labrador Sea and the evolution of the North Atlantic* in *Nature* in 1971[21] and the results of the whole cruise were published later in the *Initial Reports* in 1972.[22]

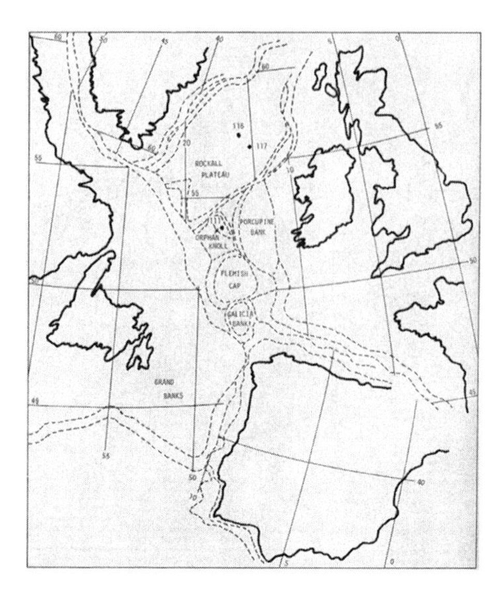

Palaeographic reconstruction of N Atlantic prior to the northward split in the Late Jurassic (150 my). Dots indicate drill sites.

The magnetic anomaly patterns of the N Atlantic consisted of stripes which could be individually dated with reference to the now-universally accepted sequence of geomagnetic reversals. This enabled me to put together a timed sequence of events for the opening of the N Atlantic. It was from the ability to date the ocean crust from the overlying sediments and from the magnetic anomaly patterns that a detailed history could be unravelled.

Following his experience on Leg 12, Whitmarsh was invited by the DSDP to become a Co-Chief Scientist on Leg 23 in 1971. He writes that

> *The drilling cruise visited both the Red Sea and the NW Indian Ocean, an area in which I had worked in 1963 with Tony Laughton and other colleagues from NIO and Cambridge. The DV Glomar Challenger was scheduled to depart from Colombo, Sri Lanka, so that there were opportunities to drill sites in the NW Indian Ocean before entering the Red Sea. Geophysical site survey information, particularly seismic reflection profiles, was scant in the area and although Leg 23 sailed with guidance from the DSDP Planning Committee as to where to drill, some sites were chosen at sea from new information gathered during the cruise, a situation very different from scientific drilling today.*
>
> *A particular problem addressed on the first half of the cruise was the acquisition of sediment cores for palaeomagnetic analysis to determine the northward drift of the Indian subcontinent during the Palaeogene.*[23] *Rob Kidd, a PhD student from Southampton University who subsequently joined NIO, sailed as the person responsible for*

extracting the palaeomagnetic samples from the sediment cores. Rob Kidd and Southampton University were in later decades to play important roles, respectively, in the future of scientific drilling internationally and in UK oceanography in general.

Estimates of basement age and magnetic data acquired during the cruise were also used to recognise new fracture zones offsetting the very clear magnetic anomaly stripes in the Arabian Sea. Joining the ship half way through the cruise was Ron Girdler from Newcastle University, a geophysicist who devoted much of his career to studying the Red Sea, and Frank Manheim, a geochemist who was to study the sulphide deposits in the hot brine Atlantis II Deep, which had been discovered by Atlantis II *and later examined by* Discovery *on return from the Indian Ocean.*

The drilling of this Deep, a hole with extremely high salinity water on the axis of the Red Sea, proved difficult. It was impossible to penetrate more than a few metres into the hard basaltic rocks, but we drilled through sediments and evaporites landward of the spreading axis and this provided important geochemical results. One site showed that, after being rifted, late Miocene salt had been warm enough to flow laterally onto younger lavas! The cruise ended in the port of Djibouti which afforded an opportunity for a geological field trip into the amazing Afar region where the three plates of Africa, Nubia and Arabia are being rifted apart on land.

David Roberts joined the Geophysics Group at NIO in 1965 having studied volcanoes in the West Indies under Prof. Malcolm Brown, and started to research the nature and origin of Rockall Plateau. Following the Penrose conference of the Geological Society of America on continental margins in 1972, he was invited to join the group charged with developing the proposal for continental margin drilling during the International Program on Ocean Drilling (IPOD). He subsequently became a member of the IPOD Passive Margin panel which developed drilling proposals for passive margins mainly in the Atlantic Ocean. His work with this panel led to the successful drilling of the margins of the Bay of Biscay, Galicia Bank and Rockall Plateau shortly after this history ends.

Subsequently several geologists and geophysicists from NIO have taken part in the DSDP and have also been involved in the management of DSDP and IPOD and its successors.

The continental margins of Europe

Rockall Plateau is one of the continental fragments which resulted from the breakup of the N Atlantic. At the northern end of this plateau, the islet of

Rockall Island itself stands up 23 m above the sea and is only 27 m wide, a granitic plug remaining from the volcanism associated with the breakup. But this is only a very small portion of the vast shoal area that covers 500 by 300 miles.

Although the general outlines of the bathymetry were known from the Admiralty navigation charts, Roberts was able to refine this and show that two shoal regions, the Hatton and Rockall Banks which run from the northeast to the southwest, are separated by a deeper area, the Hatton Rockall Basin. On a cruise in 1969, dogged by atrocious weather, he showed that this basin had a significant accumulation of sediments, sitting on rocks later proven by dredging and by shallow hard rock drilling in 1971 and 1972 to be of continental origin.[24, 25] One of the objectives of the DSDP Leg 12 in 1970, was to drill on the flanks of the basin.

Roberts spent many years studying the geological origins of Rockall Plateau and of Rockall Trough that separates the plateau from the continental shelf off Ireland and Scotland.[26] Seismic refraction profiles in the trough indicated that the crust beneath the trough, at least in the southernmost end, was oceanic in nature suggesting that seafloor spreading had initially penetrated there. This had considerable political repercussions in respect to the claims by the UK that Rockall Plateau was a continuation seawards of the continent belonging to the UK under the definitions being developed in the UN Conference (later the Convention) on the Law of the Sea. Following the initial involvements of Dr Deacon and Nic Flemming, the research of the NIO started to impinge on politics. The geophysics group was to become progressively involved in advising the government on technical issues relating to the Law of the Sea and the definition and extent of the 'continental shelf' as defined by lawyers.

Roberts expanded his interest in Rockall Plateau to that of the continental margins of northwest Europe including the Bay of Biscay, complementing the early work by Arthur Stride in cooperation with Joe Curray of Scripps. He was, in collaboration with the Institut Français de Pétrole, to bring to NIO the much more powerful industrial methods of multichannel seismic reflection profiling, with its increasingly sophisticated data processing using computers.

A new seismic exploration technique

Robert (Bob) Whitmarsh had joined NIO in 1967 from G&G, another Cambridge student of Hill. He had been developing a method of seismic refraction shooting which avoided the long upward sound path through the water which was inherent in the sonobuoy method.

By putting the seismic receivers on the ocean bottom, and firing the charges near the surface, signals from the lower layers of the crust could reach the receivers before the sound through the water arrived to obscure them. The buoys also had the great advantage of remaining in a fixed location while measurements were being made, and of experiencing lower levels of ambient noise than surface buoys. They could also directly measure, with geophone sensors, the three dimensional motion of the seafloor in response to surface shots.

This development required remotely recording buoys which could be released from the bottom by acoustic command. But the principal technological problem that had to be overcome was the provision of a suitably large instrument case that could withstand pressures of around 500 bars. Early experiments with glass hemispheres had been unsatisfactory as there were constant problems in sealing the mating surfaces. Forged hemispheres of an aluminium alloy (RR77) were designed by Dennis Gaunt at NIO and made by High Duty Alloys of Slough; some of these spheres are still in use today. The resulting so-called Pop Up Bottom Seismic (PUBS) recorders were the first of a new generation of instruments soon to be called Ocean Bottom Seismographs (OBS) and which are now widely used throughout the academic world and in the seismic industry.[27]

Three PUBS were first tried out on *Discovery* Cruise 23 in 1968 in depths up to 4,830 m, and later refined and used in many future cruises to determine oceanic crustal structure. Some of the first seismic investigations of the structure under the median valley of the Mid-Atlantic Ridge were made with these instruments.[28]

Dick Burt and John Cleverley lighting the fuse for a 150 lb seismic charge.

In 1969, Whitmarsh designed an experiment to determine whether the upper mantle beneath the oceanic crust adjacent to a spreading centre showed anisotropy in the velocity of sound waves (different velocities in different directions) which might indicate a direction of mantle flow. A PUBS was laid and shots fired to it from positions on a hexagon surrounding it. Although the experiment went very well and all shots were recorded the subsequent analysis did not indicate any variation in the upper mantle sound velocity in different azimuths. Negative experiments are often as valuable as positive ones.[29]

Launching the Pop Up Bottom Seismograph (PUBS).

Sea level changes and tectonics of the Eastern Mediterranean

Nic Flemming joined the NIO in 1967, working directly under Deacon, having started a major research project at Cambridge University on the recent variations of sea level around the western Mediterranean. He examined the raised beaches and raised remains of Roman harbours and, using his skills as an underwater diver, surveyed drowned harbours, coastal archaeological sites and underwater caves and erosion. In the eastern Mediterranean he measured sea level changes around southern Greece, the Cretan Arc, southern Turkey and Cyprus. He found that some islands had tilted during the Pleistiocene collision of Africa and Eurasia and was able to work out the marine chronology.[30, 31] In spite of a serious road accident in 1969, in which he was paralysed from the waist down, he was able to continue his research, still using his diving techniques to separate the effects of sea level changes and tectonic factors on the coasts of the islands and the surrounding continents and to show how different parts of the eastern Mediterranean reacted to the compression forces from Africa. His expertise in diving later led Flemming to develop a code of practice for UK offshore divers and he became a founding member and Honorary Secretary of the Society for Underwater Technology.

Support for the Scientific Vision

It was a characteristic of NIO that the very close liaison between the scientists and the engineers and technologists, encouraged by George Deacon, was a key to its success. Tom Tucker writes of the extremely varied demands made on the Applied Physics and Engineering groups to design and build instruments that at that time were not available to purchase, notably for the measurement and analysis of waves at sea, but also tide gauges, seismographs, ship's logs, temperature and salinity measurement and numerous other devices. Underwater acoustics plays a major part in oceanographic instrumentation and Brian McCartney, later Director of the Proudman Oceanographic Laboratory, describes the 'pingers', echosounders, side-scan sonars (including GLORIA), fish detection and improved Swallow floats. Anthony Laughton writes of the crucial importance of research ships and of life afloat. Changes in technology (including computing) enabled scientists to address increasingly complex problems at sea. Pauline Simpson, former Librarian at the Southampton Oceanography Centre, reviews how the library of the NIO, created by Jack Carruthers and further developed by Mary Swallow (formerly Mary Morgan) and by Dick Privett, was a critical resource for NIO and other scientists and later became the National Oceanographic Library. Without funding and administrative support no research would have been possible. Anthony Laughton describes how Ray Williams, Secretary of NIO, played a crucial role in serving successive directors.

16

Engineering and applied physics

'Tom' Tucker and Brian McCartney

In 1949 NIO staff were members of the Royal Naval Scientific Service (RNSS) and so could still use the excellent workshop, stores and drawing office of the Admiralty Gunnery Establishment. When we moved to Wormley in 1953 we had to set up our own facilities but because Dr Deacon believed that it was important to measure things in the sea, he ensured that our engineering facilities were of high quality and that NIO would become one of the world's foremost oceanographic instrumentation centres.

Our organisation at Teddington had been informal but the expansion of staff and facilities at Wormley required some structure. We formed an Engineering Group led by Frank Pierce and I (M.J.T.) led an Applied Physics Group (APG). This latter name distinguished it from the (mechanical) Engineering Group, and reflected the range of activities in the group; for example, underwater acoustics and applied wave research, as well as electronic engineering. The two groups worked happily together and co-ordination between Engineering and Applied Physics and the scientists was informal but effective. A key factor in this was that we went to sea together, often with members of several scientific disciplines. Thus, we all knew one another and got to understand the research scientists' interests, and they learned what techniques were available and practicable. By 1968 (the last date for which a staff list is available) the APG had nine graduates on its staff, of whom three had PhDs. The Acoustics Section of the APG was formed with Brian McCartney as its head when he joined us in 1964: he has written the sections of this chapter dealing with underwater acoustics.

NIO policy was to make any equipment it developed and built (NIO maintained its own production staff, workshop and stores organisation), available to other organisations. Since many of our instruments opened up new observing possibilities the demand was high from researchers doing research similar to ours, mostly abroad. The NIO Annual Report for 1965-66 records that between 1960 and 1965 sales averaged £25,000 per annum, (approximately £340,000 in 2010 money). The Report also noted that this activity brought our scientists and technicians into close contact with other laboratories and industry, and allowed them to build up good working relationships with small contractors: "*This activity has improved our work and lessened our own costs. It has done much to further oceanography and related studies*".

When a firm manufactured an NIO-designed instrument with only minor modifications, the Institute received a royalty. If a firm took over a design and made major modifications, NIO sold its 'know-how'. A firm designing equipment based on published information was given help and advice including sea trials. Early interactions were not always satisfactory. One firm redesigned an instrument without telling us, and another closed the department with which we had been dealing just before instruments were to be delivered. But we were on good terms with several manufacturers.

Light and radio signals penetrate only short distances in sea water so, for communications and underwater remote sensing, sound waves remain the principal information carrier. Underwater acoustics became a key technology in our work and made contributions in all NIO's scientific disciplines. Though its use in marine science and offshore applications is now widespread and commercially available this was not so early in the NIO era. Our directors and scientists nevertheless saw its great promise, and encouraged its application and development. In the 1950s and early 1960s, practically all British expertise in this subject was in Admiralty laboratories supporting the Royal Navy. A few companies, with naval contract experience, such as Kelvin Hughes and Ultra, began to produce commercial echosounders for coastal navigation and finding fish.

Underwater technology gradually transferred from the necessarily classified naval area into civilian oceanography–'swords into ploughshares'. Through the RNSS, NIO staff had good personal links with labs such as the Admiralty Research Laboratory, and the Admiralty Underwater Defence Establishment at Portland and some, such as Stuart Rusby and Jack Revie, transferred to NIO from naval work, bringing their expertise with them. Also, staff with expertise in this area were returning to universities from government laboratories (Prof. D.G. Tucker–'Tom's brother–and J.W.R. Griffiths to Birmingham, and Prof. W.D. Chesterman and P.R. Clynick to Bath). They established courses, generated research topics, published in the open literature, led conferences in underwater acoustics and the associated electronic processing, and some of their graduates joined NIO (Brian McCartney and Charles Clayson). The benefit to NIO of gifts and (often extended) loans of surplus equipment, especially transducers, from Admiralty laboratories cannot be overemphasised.

When the NIO buildings were extended in 1966 (see chapter 19), they included much-needed engineering facilities; a large workshop and stores, design and drawing offices, a wave/towing tank, pressure test vessels, an acoustic test tank and an assembly space. The 50 m long wave tank is noteworthy and its hydraulically driven towing carriage was used to test streamlined bodies and for calibrating current meters. A wave maker could, for example, produce breaking waves at a chosen observation point by programming it to run fast at first then slow down. The later, faster, low-frequency waves would catch up with the early ones and produce breakers. Windows allowed observation and photography.

Initially, our biggest pressure-test vessel worked only to 200 m equivalent depth yet we were developing equipment for oceanic depths, so in 1971 a large (1.2 m deep and 0.95 m diameter) vessel was installed for testing instruments at pressures up to the equivalent of 7,000 m depth. The new vessel was the largest in the UK and was much used by NIO and others; for example, in developing new buoyant materials enclosed in glass-fibre housings (syntactic foam).

Wave measurement and analysis

Wave research continued throughout the life of NIO, though its objectives and character changed as scientific knowledge and technology advanced. Making improved measurements of waves was a challenge. The simplest wave measurement is of changes in water level at a fixed point. To do this in the open sea, free of coastal influences, is a fundamental requirement yet poses the problem of needing a fixed reference point. The NIO approach to making such measurements resulted first in the Shipborne Wave Recorder (SBWR) and then in buoys attached by cable to a ship and finally in a big, moored data buoy (DB1) measuring waves, meteorology and currents.

In the early days, we used the piezoelectric quartz crystal pressure sensors mounted on the seabed connected by cable to recorders on shore. They were reliable and served us well from the era of Group W and throughout the life of NIO. However the charge from the crystal slowly leaked away and this meant that while it measured high-frequency (local) waves well it did not measure the longest swell and surge waves accurately. A new device was required and since no suitable commercial instrument was available in the late 1950s, we set about developing an absolute pressure gauge. Such a gauge requires a resolution of the order of 1 in 10^5 to measure long waves and 1 in 10^6 for deep-sea tides. Direct digital techniques had not then been developed, so the only possibility was a device that gave a frequency modulated (fm) output, that is, a frequency that changed with changing pressure. Frequency could be measured to high resolution and

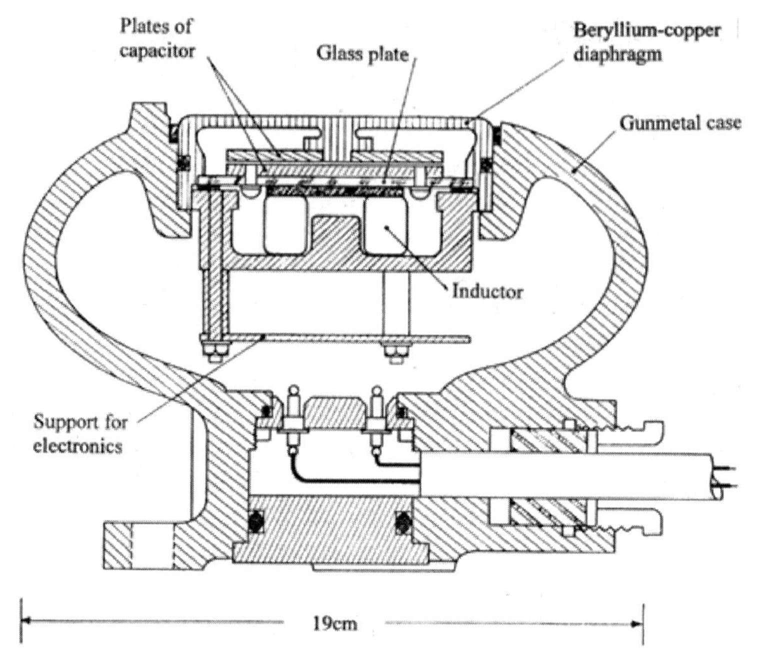

A cross-section of the NIO fm Wave Recorder sea unit.

was not affected by transmission along a cable. To get the required stability required careful design.

Referring to the figure, the pressure diaphragm was machined out of a solid beryllium copper block to give a substantial skirt with a thick glass plate supported on it carrying the lower plate of the capacitor. The capacitor formed a resonant circuit with an inductor designed for maximum stability and was maintained in oscillation by a transistor circuit. The frequency varied from 60 to 94 kHz over a pressure range equivalent to 40 m of seawater. Power was fed down a standard two-core cable and the frequency fed back up it to the shore. The inside of the device was evacuated to 0.1 atmosphere to reduce temperature effects. The overall temperature sensitivity of approximately 3 cm of seawater/°C was negligible when measuring wind waves. For long, low-amplitude waves, the whole device could be insulated or the temperature could be measured by a thermistor in an RC oscillator circuit at a different frequency. The output could be recorded digitally or converted to an analogue signal and recorded graphically. We successfully measured long waves of the order of 1 cm amplitude. These recorders served us well for many years.

In 1951, Laurie Baxter and I installed a piezo-electric wave recorder and analysis system at Casablanca where the heavy swell sometimes overtopped the wall of a new harbour. Baxter had the job of installing the wave sensor and, having done his watch-making apprenticeship

Tom Tucker testing an inverted echosounder at the reservoir.

in Switzerland, spoke fluent French. One of the crew of the local boat hired for the installation, a pleasant native African, told us he had been a slave and had once been sold for a handful of salt! The data were used in publications by Darbyshire and by R. Gelci from the Service de Physique du Globe et de Metéorologie, which had requested and paid for the installation.

The pressures under high frequency waves do not penetrate to the sea floor and so to measure the full wave spectrum, a device is needed that measures the surface elevation directly. The obvious way is a device to measure the height of the water up a pole penetrating the water surface. Such 'wave poles' had been mounted on piers in America but were difficult to install in really exposed situations. We developed the equivalent of a wave pole, a 'capacitance-wire wave recorder'. This wave recorder was simple in concept; an insulated wire was stretched vertically between two brackets, above and below the surface, the wire being looped round the bottom bracket to avoid an underwater connection. The capacitance between the water and the wire was measured and the output brought ashore by cable. We found that PVC covered wire could not be used because PVC absorbs water and water has a high dielectric constant, causing the calibration to drift by as much as 40%. Wire insulated with polythene does not show this effect.

The device was first used in an experiment to investigate the generation of wind waves on a reservoir. We used a submerged buoy moored to the reservoir bed with a vertical pole on top penetrating the water surface. The pole also carried anemometers.

Our skills with instruments had another, illustrious, user. In 1947, the Hydraulics Research Laboratory (HRL) was set up in the grounds of the NPL at Teddington. Some time before we left Teddington, Brigadier Ralph Bagnold was working at HRL. Bagnold had set up the Long-Range Desert Group in North Africa during World War II. (This Group destroyed more German and Italian aircraft on the ground in North Africa than the RAF did in the air). However, he also had an engineering degree and had produced some outstanding scientific papers on the movement of sand and sediments, and as a result of these, he had been elected a Fellow of the Royal Society. HRL was immediately opposite the back door of our lab, and Brigadier Bagnold often popped in to have a chat with us and with Dr Deacon in particular, since they were both Fellows of the Royal Society. He was a charming man and always welcome. HRL was concerned with coastal engineering and built hydraulic models of estuaries. However they had not developed all the instrumentation to monitor the models. Bagnold obviously took a wider interest in their work because he asked if we could help with the measurement of tidal height in their estuarine models. Geoff Collins and I quickly rigged up a simple device consisting of the galvanometer movement from an Evershed & Vignoles pen recorder with its long, light aluminium pen arm still attached, mounted with the axis horizontal and the arm stretched horizontally over the water. A sharpened wire was attached and bent vertically down at the end. The wire was the right length to just touch the water surface when the arm was horizontal. An alternating voltage was applied between the arm and the water. When the point was out of the water, no current passed, but as it entered the water the current quickly built up. The current was rectified and fed to the galvanometer in such a way that it lifted the pen arm, so that when properly adjusted, the point was held just touching the water surface. The current required varied with the water level and was recorded. The device worked, but we had to put salt in the water to get enough current to pass. We therefore built an amplifier for it, and then it worked beautifully. HRL and its successors used such devices for many years. They also used capacitance-wire wave recorders in their wave models.

The Shipborne Wave Recorder (SBWR)

As soon as NIO was formed, it acquired RRS *Discovery II*. I got the idea that it might be possible to measure waves using the heave of the ship combined with pressure measurements on its hull. In co-operation with Frank Pierce, this idea developed and became the SBWR, one of our most successful instruments.[1]

Two similar units are installed on opposite sides of the hull below the waterline and as near as possible to the pitch centre. Each unit has

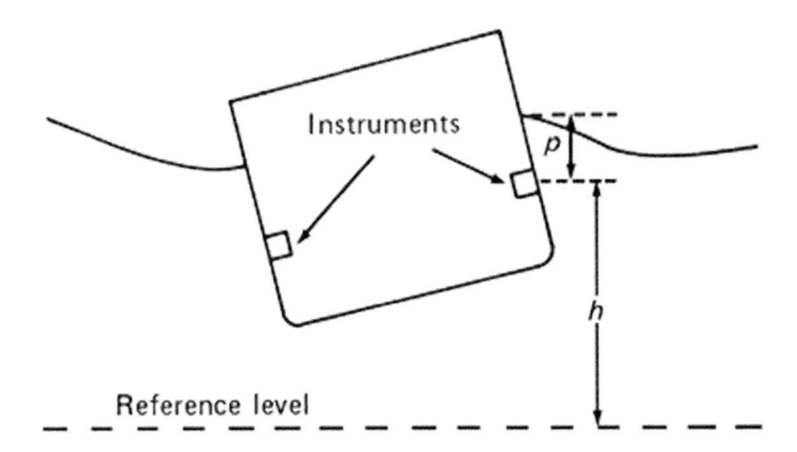

The principle of operation of the Shipborne Wave recorder (SBWR).

a vertical accelerometer mounted on a damped short-period pendulum (to keep it vertical), and a pressure gauge open to the sea. The vertical acceleration output from each unit is integrated twice electronically to give vertical displacement and is added to the pressure signal to give the height of the water surface above some arbitrary reference level. The height values from the two units are averaged to give the notional water (wave) height at the centre of the ship. This averaging is needed because if, for example, short waves approach the ship from one side, they will be reflected and doubled in height at that side, but not measured at all by the unit on the other side of the ship. The instrument does not measure waves perfectly because the ship interferes with them and also because the pendulums (more reliable and cheaper than gyroscopes) do not stay truly vertical, but it is robust and reliable. (One installed in the Seven Stones Light Vessel once gave us over 99% data return for a whole year.) Despite these limitations, the wave amplitude response was constant within ±5% for waves between 8 and 20 s and we estimated that the overall errors in wave height were within about ±10%.

The prototype SBWR on *Discovery II* went to sea for the first time in 1951. We hit a storm almost immediately, and in spite of being very seasick, the excitement of seeing a convincing wave record coming from it and showing a wave going off scale at 40 ft crest-to-trough, which was higher than any wave previously measured instrumentally, was a high-spot of my career. As data accumulated and its analysis made sense, users gained confidence in the SBWR. It could only be used in a stationary or very slow-moving ship, and so was ideally suited for Trinity House's moored light vessels and the Meteorological Office's Ocean Weather Ships. Data from these ships were used by Pierson and Moskowitz in their classic derivation of empirical equations for the wave spectrum and other

Above. *The pitch-and-roll buoy and Norman Smith who played a key role in wave buoy development.* ***Facing page***. *The Clover-leaf buoy.*

characteristics as a function of wind speed in the case where the waves have reached a state of equilibrium with the wind.[2] This is known as a Fully-Developed Sea, or Fully Arisen Sea. I should perhaps add that the existence of a fully-developed sea in the open ocean is open to question nowadays, but it is still a useful concept in certain circumstances.

Over 100 SBWRs were built and supplied all over the world.[3] The instrument, having been developed over the years, is still in use (2009) but the principle remains the same. In February 2000 a Mk 4 SBWR on *Discovery* measured a wave 29.1 m crest-to-trough which was the highest wave ever recorded up to that time.[4]

Wave direction

Around 1954 we re-started development of wave-measuring buoys, and in particular buoys for measuring wave direction. In 1946, in a remarkable unpublished paper, Norman Barber had laid down the basic principles for the analysis of the pitch, roll and heave signals from a surface-following buoy, but we did not then have the techniques to produce a practical scheme. Longuet-Higgins' experiment to measure the distribution of air pressure in a wind blowing over waves led to the development of a flat discus buoy that measured and internally recorded this, together with its pitch, roll and heave using gyroscopes. For several years David Cartwright supervised the analysis of the data. He was researching ship motion, initially in

co-operation with L.J. Rydill of the Royal Naval College, Greenwich, and later as part of a project involving the Ship Division of NPL and the British Ship Research Association (BSRA).[5] I was involved because the research needed directional wave information, but we did not know how to derive such data from the pitch, roll, and heave of buoys.

Cartwright used computers and the concept of the cross-power spectrum, not available in Barber's time, to implement a practical scheme that has formed the basis for all pitch-and-roll buoy data analysis ever since. For each wave frequency, this gives a mean direction and a measure of the spread in directions of travel. The method was tested on the records from Longuet-Higgins' buoy and gave promising results.

The digital analysis that made the computations of cross-power spectra required by Cartwright's method a practical proposition was still a formidable task. Since there were no high-level computer languages, programming required someone with a specialist degree in the subject. In 1958, we employed such a person, Diana Catton, and analysis speeded up. The records had to be read manually and punched onto cards before being fed into the English Electric Deuce computer at the nearby Royal Aircraft Establishment, Farnborough. Deuce was large and required a small army of acolytes to keep it running. It took a day to do a calculation that would now take a second or two on my own modest PC.

The analysis of the records from Longuet-Higgins' buoy thus took several years to complete. This was obviously unsatisfactory, so in 1958, in co-operation with the BSRA we negotiated with a firm called Hilger and Watts to make a recorder that took analogue signals, digitised them, and recorded them on 5-hole punched paper tape. The maximum speed of the recorder, five readings per second, was barely adequate to record the pitch-and-roll buoy signals, and not fast enough for the clover-leaf buoy. Thus, for the *Hemifusus* trials in 1964 (see below) we ordered a specially-made high-speed multi-channel digital recorder recording on punched paper tape and fast enough to record all the signals from the clover-leaf buoy. A 20-minute record from the buoy required more paper tape than was on a standard reel, so special long rolls were made for us. It was impressive to watch (and hear!).

Longuet-Higgins' buoy weighing about half a tonne was just over 1.65 m in diameter and it floated just 7.5 cm proud of the sea surface so as not to interfere with the air flow. The buoy was self-contained with its batteries and optical recorder and was kept lined up with the wind by a small sea-anchor. It was a major achievement deploying it successfully in a storm.

For experiments that did not need air pressure measurements we designed a much lighter and more stable PR buoy, that was more easily handled and recorded data aboard ship via a slack buoyant cable. It consisted of a buoyant torus 1.2 m in diameter with an instrument case mounted on a frame in its centre. A compass enabled the measured pitch and roll to be rotated to north-south axes. The analogue signals were still recorded photographically and digitised later manually. This did, however, allow the records to be examined as they were being taken and for some editing in the digitising process. It was not long before direct digital recording was used.

The PR buoys gave very coarse angular resolution, but this could be improved by roughly a factor of two by measuring the sea-surface curvature. Cartwright therefore got together with Norman Smith and Frank Pierce again and produced the Clover-leaf buoy.[6] This had three discus floats mounted on universal bearings at the corners of a triangular metal frame with a side of 2 m. Devices in the bearings of each float gave the angles of the float relative to the frame in two dimensions and the differences in the angles gave the curvature. An instrument case with the pitch, roll and heave gyros was in the centre of the frame with a compass on a short-period pendulum in a separate case above it. This buoy worked well for wave periods from 2.5 s to 17 s, corresponding to wavelengths on deep water (proportional to the square of the period) of about 10 m to 450 m. The directional resolution was about ±30°. For longer waves, the difference in slope over the distance of 2 m was small and got lost in the noise. They worked well as pitch-and-roll buoys over the whole frequency range.

These buoys were used for research in the UK and in Japan. As part of the UK ship-motion research programme mentioned above, trials were carried out on the 19,000 tonne Shell tanker *Hemifusus* in 1964 during its normal commercial voyages. Norman Smith made 21 recordings from it during brief stops on passage. The recordings, on punched paper tape, produced seven miles of error-free records. Using improved analysis programs made by Jim Crease and Brian Hinde and a faster computer meant that analysis was faster and cheaper. In Japan, Mitsuyasu and his co-workers at Kyushu University built a similar cloverleaf buoy and deployed it approximately 50 times. From these results, they developed what became one of the standard formulae for directional spread.[7]

The big advance in 1968 came from the company Datawell in the Netherlands. They produced the Waverider buoy working under contract to the Dutch Rijkswaaterstaat. This contained a clever device that carried a stabilised vertical accelerometer whose output was integrated twice to give vertical displacement and then transmitted ashore by vhf radio. Later they built a discus buoy called the Wavec that also measured pitch and roll and later still the Directional Waverider. These became effectively the world-wide standards.

The final buoy that NIO developed was the DB1, using the Wavec sensor. The project study started at NIO in January 1972, though the buoy did not get into the water until 1975. In fact, the project effectively started in 1968, when NERC set up the Standing Committee on Ocean Data Stations (SCODS) under the Chairmanship of Arthur Lee, Director of the Fisheries Laboratory at Lowestoft. SCODS had representatives of all the UK organizations with an interest in the sea and I was a member. By early 1971, SCODS had identified the need for a system of large data buoys round the UK to measure environmental conditions. NERC felt that it could not divert the necessary resources from its other work into this huge project and so passed the responsibility to the interdepartmental Committee on Marine Technology (CMT). The CMT asked the head of its Marine Technology Support Unit (MATSU), Putman, to make an assessment. He recommended that a buoy should be built as a test bed for the development of unattended instruments, telemetry systems and power supplies, and to be a prototype for the proposed network. At CMT's request NIO started a feasibility and design study on 1 January 1972.

The requirements were challenging. The network was to measure weather data, particularly to the west of the UK where it was sparse; water temperature; offshore sea level in the North Sea to help in the prediction of storm surges; and waves to provide both 'truth points' for wave prediction systems and statistical wave data for the design of offshore oil and gas platforms. The fisheries scientists also wanted a long time-series of vertical current profiles. The buoys would be designed for the

The DB1 buoy about to be towed to its deployment site.

continental shelf, but most of the sensors and systems would be suitable for deep-water use using a different hull and mooring, which would be of interest in Ministry of Defence work. The data had to be telemetered ashore in real time. It was quite a tall order. The buoy we produced met nearly all of these objectives, but because the prototype was to be used in the Western Approaches to the English Channel, it did not have to measure sea level.

To return to January 1972, Peter Collar was asked to carry out the design study, with the help of Mark Carson, Dick Dobson and Vincent Lawford of the Engineering Group, and myself. After nine months of hard work he produced an excellent 206-page Project Definition Study and an 89-page Draft Specification for firms to tender against for its construction. He and I presented these to a special sub-committee of the Ship and Marine Technology Requirements Board. We had quite a grilling, but the project was given the go-ahead with an estimated cost of £389,000 (equivalent to approximately £4 million in 2010 terms). At 7.6 m in diameter the buoy was far bigger than anything we had designed before. Rusby supervised the progress of the project with his usual drive and efficiency. Suffice to say that it was successfully deployed in the South-West Approaches in 1976, and improved versions DB2 and DB3 were built in later years.[8, 9]

The buoy as specified by Peter Collar had two unusual features. The first was its telemetry system. At that time, satellite telemetry was not yet practicable for our application. Conventional radio telemetry for digital signals gave problems for ranges beyond line-of-sight because reflection from the ionosphere caused signals to arrive with different time-delays, breaking up the digital words. We heard of a new type of radio that had been developed for the Diplomatic Wireless Service to overcome this. It was called Piccolo and used multiple frequency-shift keying (MFSK). With some adaptation it proved suitable for our use. The technique never came into widespread use because it was soon overtaken by satellite communication. The other problem was to provide a power source for long unattended operation. The Atomic Energy Research Establishment at Harwell was developing a small Stirling-engine-driven generator using propane as the heat source and capable of giving 25 Watts continuously for a year with a 25% efficiency. It had no rotating or rubbing parts; all the moving parts were mounted on an oscillating diaphragm. Peter specified this as the chosen power source, subject to satisfactory trials, but although the prototype worked well to begin with, after a month or two the burner became blocked with combustion products and the device stopped working. In the absence of a solution to this problem we had to use half a ton of batteries.

Acoustics – 'pingers', echosounders, releases and transponders (a 10 kHz family)

A capability in acoustics was fundamental to both observing and communicating below the sea surface. Much of our work focused on the selection of suitable transducers to produce underwater sound, the electronics needed to drive them and the means of displaying the signals. One of the earliest transducers from the naval surplus bin that found a very good niche at NIO was a magnetostrictive toroidal nickel scroll, used in location devices. A simple and reliable electrical circuit, originating in Group W, consisted of a capacitor charged up by the internal battery, until the voltage overcame the threshold of a Strobotron gas-filled discharge tube, which then discharged the capacitor through a winding around the nickel scroll. The scroll then resonated at its natural frequency, 10 kHz, emitting a short (1 ms) transient sound - a 'ping'. The recharging time was about a second. Their first major use was in Swallow floats. Similar 10 kHz 'pingers' were later also fitted to many instrument packages freely deployed in the sea, on moorings, on the seabed, towed by, or lowered from the ship. The free-flooded ring transducer worked at all depths, though creating the plug and socket from the instrument case to the cable winding was a nontrivial problem. (Diagrams and photographs are in Chapter 7.)

Tony Laughton with the Precision Echosounder.

Knowing the depth of the sea was a fundamental requirement for all areas of science and needed a precision echosounder (PES) capable of working at full ocean depths. Tony Laughton and John Swallow designed the first version used at NIO. The shipboard transducer for this on *Discovery II* was again found in the surplus bin, and was an array of magnetostrictive elements in a box separated from the seawater by an oil-filled diaphragm, fitted in a coffer dam, down by the keel. By serendipity, it also worked at 10 kHz, so was also able to receive and display the signals from the pingers.

The display for the echosounder and for pinger tracking was a modified Mufax (Muirhead facsimile) recorder, originally built to display radioed weather charts. Its principal feature was a linear sweep produced by a spiral conductor on an insulated drum rotating against a fixed conductor blade, with slowly advancing electrosensitive paper in between. For the PES, the outgoing ping was triggered at the start of each sweep. By varying the drum speed the display could be synchronised to one of a number of differing ping rates from multiple deployed pingers. The PES initially used electronic valves (working at a high voltage in the transmitter) but Tom Tucker and Mike Somers later replaced these electronics by safer, more flexible and more reliable transistorised designs. It became the workhorse instrument for listening to 10 kHz pings for location and telemetry, as well as for its principal role of bathymetry. The depth scale assumed an average sound speed of 1,500 m/s with a resolution of 1 m,

John Swallow (wearing glasses) working on a prototype echo-sounder fish aboard the Erika Dan *in 1958. Note the snowflakes and ice coating the ship's rail. (Photo by courtesy of Woods Hole Oceanographic Institution Archives.)*

and these depth readings were later corrected for variations in sound speed with depth and location using the Admiralty-sponsored Matthews Tables. Principal scientists on cruises deliberately chose courses across areas of ocean where there were no previous soundings to enable the cartographers to build up their charts. The Mufax record was annotated by the scientific watchkeeper and depth readings were transcribed into a logbook.

Current meter moorings and seafloor packages such as tide gauges and seismic receivers were being left for weeks and months and could be recovered, thanks to an acoustic release system developed at NIO. In the earliest versions a small spherical hydrophone capable of deep submergence received a 10 kHz fm signal from the ship. If the correct modulation was recognised a multistage relay would operate. To eliminate false operations and prevent firing during deployment, the relay needed to operate several times over a short interval before triggering the release mechanism. Early models used explosive bolts and later an electrochemical fuse (pyro) burned away a linkage to release the buoyant instrument from its anchor. Each release device was coded with one of a range of modulation frequencies, so that only the desired release would be fired if others were within acoustic range of the ship transmitter. Once the release was fired, the unit pinged away using a nickel scroll transducer at a unique preset repetition rate, to which the Mufax drum was synchronized, and the progress of the unit up to the surface and then its position could be monitored. Many anxious hours were spent watching the Mufax after transmitting a release signal, because ship navigation in the open ocean before satellite navigation was frequently in error by a mile or two and the acoustic acquisition range was only about 2 miles. A successful release was followed by a rush on deck to scan the sea surface for the surfacing buoyancy and then the instruments with their precious data were recovered.

On RRS *Discovery* in 1963, 10 kHz echosounder transducers were fitted in port and starboard coffer dams. These sites were fine in calm to moderate seas, but when rough seas caused heavy pitching, air bubbles running under and along the hull, smothered the transducer face. These bubble screens quenched the power from outgoing pulses and blanked reception of signals. To overcome this, and hence to allow operations in worse weather, NIO engineers designed a towed body to contain the transducer array, placing it about 6 m deep at 10 knots. It was not a trivial problem to design a compact body that was hydrodynamically stable, and it involved considerable experimentation in the towing tank. Towing booms were fitted each side of the new ship near the axis of pitch. The electromechanical towing cable had to be faired to reduce strumming and required specialised winches. The magnetostrictive transducer arrays were gradually replaced by piezoceramic-based ones that required less

Mac Harris (R) and Roy Wild doing final checks on a net monitor before deployment. Phil Pugh (L) is holding a transducer to transmit commands to the monitor.

maintenance. These 'Tonpilz' transducers owed much to naval expertise. Their use in NIO began with the GLORIA units, followed by 10 kHz and then 36 kHz units used in arrays. The head mass of the transducers, initially made of aluminium and later of titanium, formed part of the pressure case surrounding the unit. The ceramic transducers integrated as part of the end cap of the instrument case avoided plugs and sockets and provided an altogether more robust and acoustically efficient design. They could be used as receivers as well as transmitters, avoiding the need for the spherical hydrophone.

The last item in this 10 kHz family was the transponder: a device that emits a ping whenever it hears one, and thus looks like a big target on an echo-sounder. This had an interesting conception. Tom Tucker and Dennis Gaunt were on a cruise in the late 1960s with some biologists, and Gaunt, who was developing a winch mechanism for letting out towed nets, was having difficulty with the device for measuring the length of wire out, that is, the range to the net. It was decided that what was needed was a transponder on the net, so that the distance could be measured using the PES display. Accepting Gaunt's challenge to produce one, Tom modified one of the pingers (even winding his own 10 kHz chokes and transformers). The pinger scroll was both the receiver and the transmitter. It took three days and the transponder worked perfectly first time. The electronics were quite subtle, controlling the background noise level to just that required to give a defined statistical chance of a random firing. This transponder was probably the first made in Britain, although they became widely used in the oil industry. At NIO, they were further developed for pulse-interval telemetry, especially for monitoring biological-net functions such as depth, opening/closure, and temperature.

Improved Swallow floats

From Swallow's observations at various depths, as well as from moored current meters, it was apparent that ocean currents varied considerably on scales of 10 to 100 km and 1 to 3 months, and hence neutrally-buoyant floats would need to be detected at longer ranges and for longer periods than was possible at 10 kHz. The Mid-Ocean Dynamics Experiment (MODE) in 1972 was planned to investigate this mesoscale variability. The acoustic engineers at NIO developed a transponding Swallow float location system, called Minimode, which operated at 5-6 kHz, depths from 0-5,000 m, and ranges out to 50 km, or even longer if the float was in the SOFAR (Sound Fixing And Ranging) channel around 1,000 m depth (it was a natural acoustic wave-guide).[10] The low frequency was chosen to give the range required and was made possible by the availability of efficient wide-band ceramic ring transducers that were ideal for the purpose.

The recovery of a Minimode float. These could be tracked at ranges up to 100 km.

Up to 18 floats could be tracked at once by virtue of simple frequency coding. At these ranges, it was necessary to convert acoustic travel times to range by correcting for the refractive propagation, a process made much easier by the presence of the onboard computer. The sound velocity vs. depth profile needed for this was calculated from standard Conductivity-Temperature-Depth(CTD)profilerdata. Each float was fitted with an acoustic release so that a ballast weight could be dropped, the float retrieved and recycled before battery life was fully expended. Batteries lasted several weeks since the transponder only emitted when the ship transmitted to it for a fix. It was good fun seeing the float tracks from different depths build up on the plot, and then recovering the floats and turning them round for more tracks. American colleagues went to an even longer-range system using 600 Hz organ-pipe transducers, though the longer ranges restricted the floats to a narrower depth range around the SOFAR channel axis (500 m to 1,500 m, depending on location).

Tide gauges

Traditional tide measurements used 'stilling wells' built into harbour walls and piers, but if you wanted to record tides where there was no suitable pier, the only practicable option was to record pressure on the seabed. In the early days of NIO, the power requirements and the bulk of the electronics were too great for a self-contained system, and in any case we did not have the necessary recovery systems. In 1959, Malcolm ('Mac') Harris started developing a system with a measuring unit on the sea-bed attached to a surface buoy transmitting the data ashore by radio. It used the American 'Vibrotron' sensor; this had a wire maintained in vibration, and whose tension, and thus its vibration frequency, was varied by the external pressure.

The first application was in 1962: the Mersey Dock and Harbour Board wanted to know the exact variations in depth of water over the Liverpool Bar. The radio signals were received at the University of Liverpool Tidal Institute (ULTI) at Bidston, from the site in 11 m of water a few kilometres away. After some teething problems, in particular with radio interference, several weeks of good data were obtained before the cable was cut by a ship that had strayed off course. Though the instrument was repaired and in 1964 transferred to ULTI, it was clear that the system would always be vulnerable.

The tide gauge being deployed. The fm pressure transducer can be seen at the bottom of the frame.

By the late 1960s very low-power CMOS-integrated circuits were available and these allowed instrument packages of greatly increased sophistication that could be left unattended for long periods. An early application was the NIO recording tide gauge for use on the continental shelf. This instrument was developed by Peter Collar at about the same time as a deep-ocean tide gauge was built by Frank Snodgrass at SIO. The NIO development was to provide data to support Cartwright's work on the theory of ocean tides in the North Atlantic.

Initially, the instrument had to be capable of operating for a lunar month so as to allow separation of the principal lunar and solar components. Having proved the reliability of the instrument the intention was to deploy tide gauges at 100 km intervals along the continental shelf edge between the South West Approaches and the Norwegian Shelf so as to define the boundary conditions for both oceanic and continental shelf tides.

The offshore tide gauge used the reliable fm sensor but with a stiffer diaphragm to enable it to be used in depths of up to 200 m. The data were recorded in digital form on magnetic tape. The prototype was used for the first time in early June 1969. The initial trials were not without incident, as described by Collar:

> *The sea is a hard taskmaster and the way towards achievement of reliability in new instrumentation is beset with pitfalls for the unwary. The site chosen for initial testing was the Hurd*

> *Deep, a long depression in the seabed providing a water depth of a little more than 100 metres in the English Channel to the northwest of Guernsey. All seemed well when the tide gauge was laid on the seabed under the watchful eye of John Swallow. Sea conditions were almost ideal for the initial 24 hour deployment. Just for this one occasion a recovery line buoyed up with surface floats was attached to the seabed capsule for ease of retrieval and to provide a back up for the new type of thermally activated release that was being tested. On this first occasion, however, many anxious hours were spent trying to recover the tide gauge, because the fuse for the thermal release had failed to fire. Even worse, the surface floats had become detached from the recovery line, rendering it useless. The technique now resorted to was dragging. A length of several hundred metres of 8 mm wire equipped with heavy chain and grapnels was repeatedly dragged round and across the position of the reluctant tide gauge.*
>
> *This, too, proved unsuccessful and operations had to be suspended overnight. In those days satellite navigation of any sort still lay in the distant future and navigation in Continental Shelf waters relied mainly on the hyperbolic Decca chain or Loran C system. The drawback was that ionospheric radio propagation effects rendered the ship's Decca system unusable during the hours of darkness. Although work began again at first light using a more complex dragging arrangement, it was not until late in the evening on the following day that the tide gauge was recovered, to the relief of all, none the worse for some rough handling. It was then redeployed for a further 15 days and its subsequent recovery was entirely straightforward.*

Various improvements were made to the original instrument. One undesired feature–significantly improved upon but not eliminated until proprietary quartz crystal transducers became available–was a slow drift in the output of the pressure measurement transducer. For tidal measurement this was not serious but it precluded accurate measurement of slow changes in mean sea level over periods of more than a few days.

Thanks to some sterling work by Bob Spencer, Pat Gwilliam and Bob Kirk, the programme of shelf edge deployments was completed by 1973. The work undoubtedly provided the springboard for the subsequent development at IOS Bidston (to which the work on tides was transferred) of deep oceanic tide gauges. With greatly extended endurance and stability these gauges nowadays permit measurement of sea level changes over periods of up to two years.

Seismographs

A seismograph record usually shows small oscillations of the earth with a period of a few seconds, called microseisms. Early studies of these showed that they are generated by sea waves and have half the wave period, but did not manage to identify where they were generated. If this could be done, it could potentially be valuable in wave prediction. Dr Deacon had written our first paper on the subject and retained a keen interest in it. Thus, when H.M. Iyer of the Indian Naval Physical Laboratory at Cochin came to us with a two-year UNESCO Fellowship, Deacon suggested that he study this problem with a view to obtaining his PhD. This he did with the co-operation of Jack Darbyshire and the assistance of Brian Hinde who continued with the project when Iyer went home. Initially the only seismographs available to us were the Galitzin instruments at Kew Observatory. These were mechanical devices magnifying the ground movements by a factor of 600, so that microseisems were only a few millimeters in amplitude on the records and had to be enlarged and then digitised by hand. We then designed our own seismographs with an electrical output and installed them in two vaults, one in the grounds of NIO, and the other 5 km (approximately half a microseism wavelength) to the south in Haslemere. This was connected to NIO by Post Office landline. All the signals were recorded in both analogue and digital form, and the installation was in full operation for over a year, recording all six components. From the instrumental point of view, this installation was very successful and we were proud of it, but the structure of the microseisems proved so complicated that we could not get useful information about the direction of propagation.

The principle of the seismographs was novel. They were accelerometers whose outputs were integrated electronically (as in the SBWR). The horizontal one had to measure accelerations equivalent to ground tilts of the order of 10^{-7} radian (approximately 2 arc-seconds) and the vertical one changes in the apparent weight of a mass of the order of 1 part in 10^7. They had to be both compact and robust (we could not afford to build conventional vaults, effectively large underground rooms) and capable of untended operation.[11] We sold several of these seismographs to other organisations and, although we failed to measure the direction of travel of the microseisems, we learned a lot about their propagation and our results were useful to seismologists studying in this field.[12]

Electromagnetic (EM) current meter and ship's log

Measuring water flow is traditionally done using an impeller or rotor for speed and a vane and compass for direction, but for rapidly fluctuating currents such as those due to waves or turbulence, this technique is not

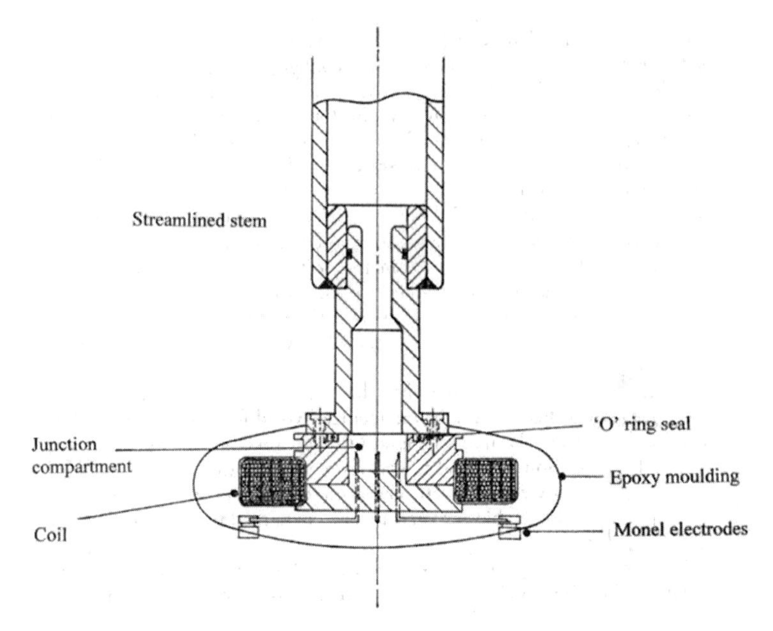

The ship's log version of the electromagnetic current meter.

suitable because of the slow sensor response. For a conventional ship's log only the forward (or astern) speed is measured even though the ship may be moving sideways. To address both of these problems we developed an instrument based on electromagnetic induction. The principle is simple. A magnetic field is produced by an electric current flowing through a coil of wire in an insulating streamlined ellipsoidal (mushroom-shaped) body. The seawater (an electrical conductor) flowing through the magnetic field produces a voltage at right angles to both the field and the current, and this is detected by two pairs of electrodes on the surface of the ellipsoid.

The first instrument based on this principle (the ellipsoid was about 40 cm in diameter and 5 cm thick) was made by Norman Barber in Group W. We quickly discovered many problems. The principal one was that one could not use direct current in the coil because, however hard we tried to avoid them, there were always electro-chemical emfs (voltages) of at least the same order of magnitude as those we were trying to measure. Thus, we had to use alternating current (AC) (for practical reasons, the 50 Hz mains), but the AC magnetic field induced voltages in the leads to the electrodes that were much larger than those we were trying to measure and we were unable to balance them out sufficiently well, so the project was shelved.

Some 10 years later, Ken Bowden, then at Liverpool University, wished to measure turbulence in a tidal stream in relation to the movement of seabed sediments, and the EM current meter, with no moving parts, seemed ideal. Using the improved electronic techniques then available, we made an instrument that met his needs: crucially he was looking at fluctuations in the flow, and so zero stability was not too important. He was able to measure the turbulent components in both the horizontal and vertical directions, and thus to calculate the stress on the seabed.

The computerised navigation system developed by Jim Crease for *Discovery* in 1969 calculated the ship's position between occasional satellite fixes from its movement through the water (dead reckoning). Since the ship spent a lot of its time stationary or moving slowly, the athwart-ship movement due to the wind could not be neglected and needed to be measured. The two-component EM current meter was ideal but in this application the zero stability was important. We had found when calibrating the earlier instruments that any nearby obstruction in the water caused large spurious signals. This was because the device acted as a sort of transformer, inducing large emfs (voltages) in the water even when it was stationary. If everything were axisymmetric, these are balanced by the effect of the resulting electric currents, but an obstruction disturbs this balance and large voltages appear.

The answer was to use switched DC, waiting for the transients to die down after each reversal before sampling the voltage. The result was satisfactory, giving an accuracy of 0.1 knot or ±1% for the forward component whichever is the greater and 0.2 knot or ±2% for the athwart-ship component. The measuring head was mounted on a streamlined spar that was housed just inside the hull in port, and lowered to place the head 1 m from the hull at sea.[13]

This instrument was later used extensively as a current meter. One of these applications was to measure turbulence near the deep ocean floor at depths between 900 m and 1,600 m in the Gulf of Cadiz: and where, as the result of a failed release, the instrument probably still remains.

Temperature and salinity

From the time of the mid 1960s continuous profiling of temperature and conductivity (and hence salinity) pioneered by Neil Brown and Bruce Hamon in Australia had replaced traditional water sampling using bottles and thermometers.

In about 1960, there was a need for a similar instrument of lower accuracy that could be used in coastal waters and particularly in estuaries where salinity contrasts were large and boundaries between fresh and saline water were sharp. Hamon had designed an instrument for this application[14] that Brown had later transistorised, and Hamon asked if we could get it into commercial production. He had noticed that the variation of resistance of a thermistor with temperature was rather similar to that of the resistivity of seawater. Thus, if one made a Wheatstone bridge with a conductivity

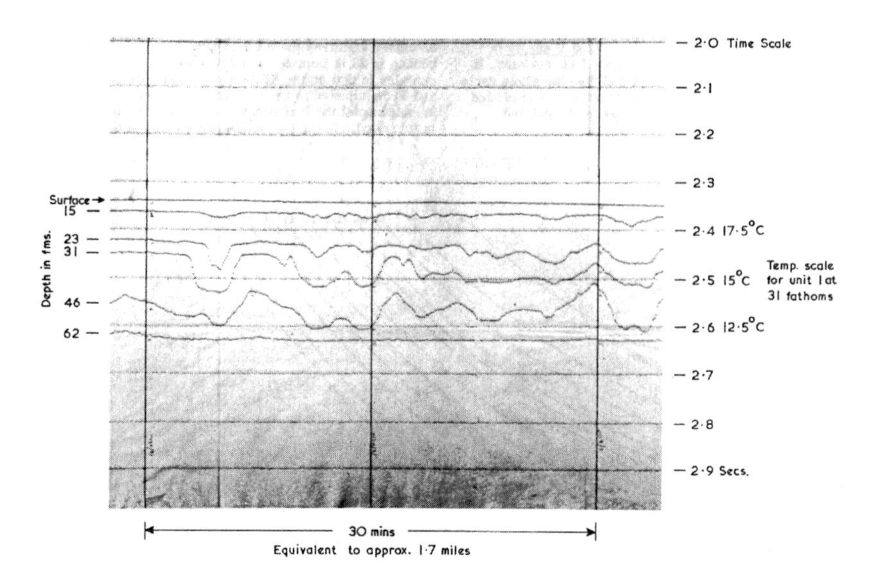

Thermistor chain record near the continental shelf edge. The waves are confined to the seasonal thermocline between 20 and 60 fathoms (about 40 to 120 m). The thermistor sensors were sampled every 5 seconds.

cell in one arm and a thermistor in the other, temperature effects could be balanced out. With some 'tweaking' of the thermistor characteristics an accuracy of 0.06 in salinity and 0.1°C could be achieved over a wide range of salinity; adequate for most estuarine studies. The sensors were on 100 m of permanently attached cable. John Moorey built one to the Hamon/Brown transistorised design and it worked well, so we found a firm who wanted to make and market them. The performance of the commercial prototype was hopelessly bad and we found that they had redesigned the instrument. The key to our success was choosing the 'tweaking' resistors by trial and error and calibrating the dial for that particular instrument. Such an approach was not commercially viable for a production instrument, so they had redesigned the instrument without consulting us. We then devised a systematic way to choose the resistors and to make all the dials the same and after making a number of instruments ourselves to ensure that no further problems arose, it went into successful commercial production. After this experience our commercial agreements contained a clause that no changes to the design were to be made without our agreement.

To investigate internal waves, we needed instruments that could measure both the vertical and horizontal temperature structure. The 'thermistor chains' developed elsewhere to do this carried a series of thermistor thermometers on a vertical wire but required a separate conductor in the cable for each thermometer. This was a serious limitation and in the mid 1960s Roy Bowers and Derek Bishop developed a clever system that

The Helical Warping Capstan with a thermistor module on the traction unit. The storage drum is in the background. The rubber fairing on the cable stops the tow cable vibrating (strumming) as it is dragged through the water.

required only one conductor in the cable to telemeter all the temperature and depth signals.[15] A considerable number of units could be attached to the cable. The accuracy of measurement was 0.02° C. Data were recorded on a Mufax facsimile recorder.

Since the cable needed to be towed as nearly vertical as possible, it had a streamlined weight of just under a tonne at its bottom end, and had a rubber 'fairing' clipped to the cable to streamline it. This could not be stored under full tension on a winch drum without damaging the fairing, so a capstan was required to pull it in so that it could be stored under low tension on a separate storage drum. Now a conventional capstan is a simple rotating drum flared to the edges. The wire is wound round it two or three times and a gentle pull on the inboard end tightens the wire and pulls it in. The turns are wound in at the edges and slip to the middle to make room for the next turn. It is not a calamity if they sometimes ride on top of one another. In the case of the thermistor chain, however, this system would damage the fairing. So Frank Pierce designed a 'Helical warping capstan' that overcame this by having two interlocking drums, each consisting of a flange with extremely strong cantilevered fingers interlocking with those of the other drum. The axes of the drums were offset and at an angle to one another and arranged so that as they both rotated at the same speed, the second drum lifts the cable off the first and moves it sideways by a few cm before re-depositing back on the first. This happens at each rotation, so that the cable lies as a helix on the capstan and does not overlap itself. I believe that capstans working on this principle were in use in the textile industry, but it took great skill to translate the idea into a working design for our purpose. Later, Denis Gaunt developed a double-barrelled capstan that performed a similar function but with simpler mechanics.

Side-scan sonar and GLORIA

As described by Stride in chapter 14, side-scan sonar, which displays echoes from the seabed, became a major tool for the NIO's geologists. The first NIO installation, designed by Tucker, Stubbs and Pierce[16], used ex-RN 36 kHz transducers in an array fitted outside the hull of *Discovery II*. (Figure in chapter 14). The array, transmitting a 1 ms pulse, gave a narrow (1.5 degrees) beam in the horizontal plane; the beam in the vertical plane, about 30 degrees wide, was directed obliquely downwards by an adjustable angle. A range up to about 1 km was achieved in continental shelf depths of 200 m or less. As the ship steamed forward the beam insonified a strip of seabed at right angles to the track. The sonar receiver fed a Mufax display, creating a line whose intensity represented the backscattered signal from the strip. Multiple strips side by side built up a view of the various geological formations and sediments.[17] Industry later produced side-scan sonars, generally at higher frequency, shorter range and higher resolution, for wreck location and identification, site surveys for production platforms and other marine engineering purposes.

Due to the roll of the ship, it was important to stabilise the transducer array, and this was achieved with the expertise of Jack Henderson, an electrical engineer from Birmingham University, using a metadyne automatic control system referenced to a vertical seeking gyroscope, located near the ship's centre of roll. Blocking by bubbles under the hull still limited operations in heavy pitching conditions.

This stabilised 36 kHz transducer array was sectionalized, so that separate elements along the axis could be cabled up into the laboratory, enabling other engineers from Birmingham to experiment on 'within pulse' electronic scanning for fish detection. The junction box for the cabling of these separated elements of the transducer was situated just inside the hull, right down under one of the boilers and just above the bilge. Wiring up the cables to this was not a favourite occupation, as the author remembers well when a gasket leak caused the junction box to fill with bilge water, ruining scanning experiments for a day or two. As these trials were in the late 1950s, second-hand wartime electronic valves were used for cheapness, but sadly proved too microphonic for shipboard use. At the next call into a port a trip to a radio shop purchased modern valves and the system was redesigned overnight and built in two days. Resources were found for a transistorised design once back at Birmingham. It was a valuable lesson to learn, that when ship time costs so much, short cuts and economies that compromise reliability are not wise.

The new RRS *Discovery* had a special floodable sonar trunk, whereby a metre square section of the ship's hull bottom could be replaced by a plate fitted with a sonar pod, in which different transducers pointing to port and starboard could be mounted. These arrays, about a metre below the hull, largely avoided turbulence and bubbles so that good records were obtained even in poor weather. A more modern hydraulic servo system set to stabilise at the selected declination angle was designed and fitted by the NIO engineers.

As well as the 36 kHz side scan, a shorter range (300 m) system at 250 kHz was designed and built to give the geologists finer resolution in inshore waters. This frequency was also found to be useful for the study of near-shore and near-surface hydrodynamics. Stationary side-scan transducers were positioned near a beach looking seawards and the data cabled ashore and logged. Breaking waves, bubbles and other near-surface processes were studied. Longshore currents were studied using the sonar pointing along the beach.

After the success of the 36 kHz side-scan sonar for continental shelf regions, it was natural to look at scaling up a side-scan sonar for the deep-ocean sea floor. An acoustic feasibility study by Ron Stubbs showed that longer ranges (30 km, say) needed a lower frequency (10 kHz or less), a larger array and more power. The concept of Geological Long Range Inclined Asdic (GLORIA) was born. In 1965, Dr Deacon was persuaded that the large resources required were justified and the project began, headed by Stuart Rusby, with Mike Somers and Jack Revie of the Applied Physics Group, and Frank Pierce, Dick Dobson and Roger Edge of the Engineering Group. GLORIA was the largest technology project handled by NIO.

The choice of 6.5 kHz ensured that the range could be achieved, and that the power could be handled by the transducer without cavitation. Signal processing was also a complex issue. Though a Deltic (Delay Line TIme Compressor) signal processor became available, NIO did not have any experience with it and so the system was also designed with a simple 10 mS monopulse at large (60 kW) peak power. The Deltic was not used and later a linear magnetic drum correlator became available from ARL, so that a long wide-band (linear frequency sweep) pulse could be transmitted at lower peak power, but the same overall energy. A tremendous amount of effort also went into the transducer design and reliability and a deep (flooded) disused salt mine shaft was used to test the transducers to their limits.

GLORIA Mk 1 trials cruise in 1969 featuring the vehicle on its davit, and many of the individuals involved in its development. (Left to right)
Seated front: *Ray Peters, Dick Dobson, Stuart Bicknell*
Middle: *John Swallow, Ships Officer, Norman Smith, Harry Moreton (bosun), Dick Burt (netman), Ships officer, Capt. Geoffrey Howe, ??, Stuart Rusby, Mike Somers, Brian McCartney*
Back: *Brian Barrow, Vince Lawford, Keith Tipping, Stuart Willis, Roger Edge, Percy Woods.*

To achieve the 10 km range in the deep ocean, the transducer had to be towed at a depth of about 50 m to avoid the sound being bent downwards in the warm, high-sound-velocity surface layers. The higher pressure at that depth also had the advantage of reducing cavitation, minimising surface echo effects and giving some decoupling from the ship's motion. The initial transducer housing was a towed body 10 m long, 2 m in diameter and weighing 6 tonnes. To handle this on *Discovery*, one of the lifeboats had to be removed and replaced with special davits to lower and recover the vehicle.

The project had two early setbacks. Firstly, it was thought that so valuable a vehicle towed in open ocean conditions presented an unacceptable risk without some back-up safety system. Almost a year was spent trying to design a buoyancy recovery scheme that would operate if the cable broke. It turned out that to do so would require such a rapid generation of gas in the sinking vehicle, similar to that in rocket propulsion, that the safety system would itself be dangerous and immensely expensive.

Secondly, during trials of the handling system in Southampton, one of the davits failed and the vehicle dropped into the water; luckily the man on top of it was uninjured. It took months to repair the damaged vehicle.

Launching in the open ocean required a team of technicians in an inflatable dinghy and divers to secure and release the handling gear. Tom Tucker was on board for the first sea trials and got worried about the safety of the divers, who were undertaking some risky tasks. When he got home, he asked Dr Deacon if Nic Flemming (an experienced diver and a very active member of the British Sub Aqua Club) could prepare a code of practice for NIO divers to ensure their safety. Deacon agreed and the code of practice was soon implemented and was also adopted by the British Sub-Aqua Club and throughout NERC. Eventually, it became the starting point for the Government's code of practice for UK offshore professional divers. Out of acorns, mighty oak trees grow!

Initially GLORIA was towed using an armoured cable that was nearly vertical down to the vehicle. The cable needed a streamlined plastic fairing to prevent cable vibration and the vertical wire transmitted the vertical motion of the ship to the vehicle. A scheme was devised to decouple the vehicle from the ship. The towline (an armoured electrical cable a few centimetres in diameter) was linked to an accumulator system of hundreds of metres of stretchy rope running through pulleys on both sides of the ship. The electrical cable looped across the afterdeck hung on an overhead wire. Although it worked well, it was a real 'cat's cradle' and very dangerous to be near.

GLORIA was first used by Arthur Stride in the Mediterranean on *Discovery* Cruise 29 in August 1969, and later on that cruise Tony Laughton used it in the open Atlantic. Although four days of sonographs of the Azores-Gibraltar plate boundary were obtained, the launch, recovery and operating procedures proved to be very unsatisfactory in the rougher Atlantic conditions. A new towing arrangement was clearly needed. In the new scheme the cable led directly to a separate streamlined float below the depth of wave action and well astern of the ship. From this the heavy GLORIA vehicle was suspended at an appropriate depth. The float/vehicle combination was nearly neutrally buoyant and effectively decoupled the vehicle from wave motion. The dangerous accumulator ropes were not needed. This system was used successfully with GLORIA Mk I for six years.

Eventually it was decided that that a Mk II version of GLORIA should be designed and built. The brief for Mk II was that it should be transportable between ships, require no personnel or boats in the water and take minutes rather than hours to deploy, be two-sided in operation and require a sea-going team of no more than three. Tests with Mk I had demonstrated that less directivity in the vertical plane would be preferable and this allowed a reduction in the diameter of the towed body. Mk I required an enormous electrical supply housed in a special room on *Discovery* to power the transmitter valves, each some two feet high! Since signal processing became available the lower peak power for Mk II could be provided by a solid-state transmitter housed in standard instrument racks. GLORIA Mk II was used successfully for many years, and in particular we obtained a very successful commission to survey the entire US Exclusive Economic Zone, but the story of this will have to wait until the history of IOS is written.

Along with the PES and the side-scan sonar, the geologists and geophysicists used seismic profiling to examine the vertical structure under the seabed. Initially, NIO engineers designed and built these systems; boomers (low-frequency electrodynamic sound sources), sparkers (electrical discharges) and air-gun sources and long (100 m) towed hydrophone arrays. However, bigger and better systems designed for the hydrocarbon exploration industry soon became commercially available and were used.

Fish detection

The Development Commission was keen to use NIO's underwater acoustics expertise in fish-detection research. Actually the problem was not one of fish detection but of species identification and sizing of the fish. In collaboration with D.E. Weston of ARL and F.R. Harden-Jones of the Fisheries Research Laboratory, Lowestoft, the Birmingham sector scanning sonar was used to detect the depths and shape of pilchard shoals off the Cornish coast, and the hypothesis that the pilchard shoals were good targets of the 1 kHz coastal-based sonar frequency because of swim bladder resonance was postulated.[18] NIO began a programme to examine the target strength of fish, and in particular the phenomenon of swim-bladder resonance.[19] Not all fish have swim bladders, but those that do have an enhanced acoustic target strength. The possibility of an even higher reflection at a resonance frequency followed from the known acoustic resonance of air bubbles in water. Theoretical calculations suggested that to first order the resonant frequency would relate to fish size (inversely proportional to the length) and depth (inversely proportional to the square root of depth), and to second order on bladder shape and condition. For fish of commercial sizes, e.g. of cod, haddock, herring at continental shelf depths, resonant frequencies were expected in the range 100 Hz to 2 kHz. For smaller fish, e.g. myctophids at oceanic depths, resonant frequencies could be as high as 36 kHz.

With the diving assistance from fisheries biologists from the Marine Laboratory, Aberdeen, the resonance theory was validated for live fish in experiments from FRV *Mara* in Loch Torridon. The target strengths of several species and sizes of fish were measured over a wide frequency range. The resonances were highly damped, but still gave much enhanced target strength over a wide band. In these experiments on *Mara*, live fish were maintained in a large mid-water cage hanging down at one end of the ship. Divers took one of the fish and transferred it to a smaller experimental, acoustically transparent cage attached to another weighted vertical rope. This rope was then walked along the ship's side to the operating station at the other end of the ship. Not many people can claim to have taken a live fish for a walk!

Generating low-frequency sound in water is inefficient and non-directional with modest sized transducers (which is why seismic sources are usually explosives or air-guns) and so resonant sizing of commercial fish would be uneconomic for commercial fishermen. Target strengths of fish at higher frequencies are highly dependent on the orientation of the fish to the sonar or echosounder beam and are also complicated by interactions within shoals. The work was not continued at NIO, but NERC funded a university programme to continue the target-strength work.

In September to December 1965 the SOND cruise set out to examine the vertical distribution of oceanic fauna in mid Atlantic. The NIO towed sampling nets were opened or closed by acoustic command from the ship so that the depth horizons of the catch were well established. They also carried an acoustic (10 kHz) telemetry system to monitor the depth, temperature and open/closed state. This cruise was an opportunity for the acoustic scientists to examine simultaneously the deep scattering layers (DSL) observed on echosounders; a phenomenon first reported by Naval ships in the 1930s. The NIO team used the 10 kHz PES, the 36 kHz sonar directed down as a sounder (also this could be operated at an alternative 54 kHz), and a 250 kHz sounder. Scattering layers were observed at different depths at different frequencies, many of the layers migrated towards the surface at night and many of the small fish caught had swim bladders, which could explain the frequency selectivity. An attempt was made to calculate the backscattering strength of the catch and relate this to the volume scattering depth profile, most successfully at 10kHz, less so at the other frequencies for which calibration at sea was more difficult.[20] Fish are not the only fauna with gas enclosures in the oceanic plankton; for example, siphonophores have strings of small bladders of gas.

Miscellaneous projects and devices

Lt. Cdr Brian Bary of the Royal New Zealand Navy came to work with us for two years in 1952. At that time there was interest in using electric fields to guide fish into nets; scientifically, it might enable us to catch some of the more active species that escaped ordinary trawl nets. From a commercial standpoint, I (M.J.T.) met a German entrepreneur who hoped that he could use electric fields to guide fish into the mouth of a sort of underwater vacuum cleaner. Bary set about getting some precise knowledge of these effects. Because fish are vertebrates, experiments on live fish had to be done under supervision at a suitably licenced establishment. Fortunately, the RN Physiological Laboratory at Alverstoke, near Gosport, had such a licence for its work on the effects of diving, and agreed to let Bary do his experiments there. He used a perspex tank 180 cm long and 44 cm square filled with filtered seawater. He used mullet as near as possible to the 'standard' length of 20 cm.

Bary experimented with AC and DC fields produce from the AC mains and batteries respectively, though the power required for these in the full scale would make them impracticable. He also used pulsed electric fields for which I designed a generator with the pulse length adjustable from 95 to 9,000 microseconds (μs) and a variable repetition rate. I think the peak power was 500 W. Brian found that the most effective pulse duration was 500 μs matched to an optimum frequency of about 30 Hz. This stunned his 'Standard Mullet'.[21]

In 1950 we were approached by Professor J.A. Steers, of the Department of Geography at Cambridge, who was interested in studying how the shape of a beach as one moves from land to sea changed during storms. The usual surveying methods were too slow to provide all the data they wanted, and could not be used much beyond the water line. Laurie Baxter designed two rather elegant instruments for them. The first was like a perambulator that you pushed down the beach. Inside was a damped pendulum whose deflection was recorded on a paper chart driven by the tracks (later large wheels), thus giving a record of the slope. This 'beach gradient recorder' allowed one to fill in detail between conventionally surveyed heights. The second, the 'underwater profile recorder', resembled a small garden roller. The depth was recorded by the pressure of the water on a metal bellows that moved a stylus across paper tape 2.5 cm wide. The paper was driven by a gear train operated by a pendulum that remained vertical as the roller turned: if the roller turned over, the paper kept going in the same direction! It was taken out to sea to a maximum depth of 5 m in an inflatable and then pulled ashore. It operated with an accuracy of about 15 cm. I think that this was found to be the more useful of the two instruments.

In 1958, Ron Currie was studying the factors governing the productivity of the oceans, and wished to measure how the total amount of light available to the phytoplankton for photosynthesis over, say, a 24 hr period, varied with depth. Laurie Draper worked with him to develop a self-contained integrating irradiance meter. The photocell consisted of an evacuated glass enclosure containing a cathode that emitted an electron whenever a photon of light fell on it. An anode collected these electrons. The resulting current was accumulated on a capacitor and each time the voltage reached a certain level, an electronic circuit discharged it and operated a mechanical counter. Thus, the number of counts over the deployment period gave the amount of light that had fallen on the device. The instrument was contained in a tube with windows at both ends: one had the photocell behind it, and the other the message register. When it was stored upside down, mercury relays switched it off. It was a satisfyingly simple and effective device.

Dr Jack Carruthers, the high priest of low technology, devised several simple but ingenious instruments. His famous jelly bottles could measure the inclination of trawl wires and currents. A bottle was partly filled with hot (and therefore liquid) jelly. It was then attached parallel to the trawl wire as it was deployed. While trawling, the jelly set and preserved the slope. For current measurement, a compass card was floated on the jelly. The buoyant bottle joined to an anchor by a string tied to its neck was lowered to the seabed and left. The bottle would tilt with the current and the setting jelly preserved the angle (current speed) and orientation (direction). Carruthers was an enthusiast for these devices and according to my count, published 24 papers about them between 1954 and 1972. I will quote just two here: one[22] gives a good account of the earlier devices the other[23] gives an account of his last version and a few references to earlier papers.

'J.N.' Carruthers proudly demonstrates on of his devices assisted by 'Dick' Privett. NIO transport department in the background.

Postscript

The NIO Annual Reports document the key role played by the Applied Physics and Engineering Groups in supporting scientists with innovative solutions to practical problems. At every stage, they underpinned NIO's science. The close working relationships between staff was a key element and was cemented both by the shared experiences and rigours of seagoing and the lack of 'stratification' within the Wormley lab as exemplified by the almost universal use of the canteen.

From a 21st-century perspective, many of the devices will seem primitive and fit now for museum display but they incorporated clever mechanical design and simple, robust state-of-the-art electronics. Sadly, few examples survive as a great deal of obsolete equipment was scrapped when space at Wormley was at a premium in the 1980s.

17

Research ships

Anthony Laughton

For many oceanographers the most important tool is a research ship on which to go to sea to collect data and to test theories. Much can be done in the laboratory developing theories on the basis of physics, chemistry, biology or geology, but the hard facts determine whether the theories will stand the test of time. Today, with the advent of satellites and remote sensing, it is still essential to 'sea-truth' remotely sensed data by making *in situ* observations, and of course satellites only observe the very surface of the ocean.

NIO was fortunate, being created partly from the Discovery Investigations, to have two research ships, which had been, before the war, the workhorses of the whaling investigations in the Antarctic. Both RRS *William Scoresby*, a whale catcher/trawler of 324 tons, and RRS *Discovery II*, built for research, had wartime careers followed by major refits to prepare them for the research programmes of NIO. The *William Scoresby* was used for the 1950 study of the Benguela upwelling system and some associated whale marking but on her return home she was broken up and played no further part in the history of NIO.

Discovery II (affectionately called *DII* hereafter and pronounced D2) was specifically built for research in 1929 when the Discovery Investigations realised that their former research ship, the *Discovery* of Captain Scott fame, was unsuitable for work both in the ice and in the stormier waters of the Southern Ocean. *DII*, built by Ferguson Bros (Port Glasgow), was 234 ft long with a 36 ft beam and a draft of 36 ft when fully loaded and of 1,036 tons gross. She was propelled by a triple expansion reciprocating steam engine fed from an oil-fired boiler. This arrangement generated smooth, quiet and flexible power with little vibration.

Because of the risks of getting squeezed in the ice pack, her bottom profile was almost cylindrical and this meant that she rolled hugely even in moderate sea conditions. Later two longitudinal keelsons were added to reduce rolling. These were designed so that they could be torn off without

RRS Discovery II *dressed over all.*

damage to the hull if she was caught in ice. In 1950, His Majesty the King approved the prefix of Royal to the Research Ship and the Admiralty granted a warrant for her to use the Blue Ensign 'undefaced', a privilege not often granted.

Following her post-war refit her first cruise was the 'long commission' to the Antarctic in 1950-52 described in the Ocean Ecology chapter. This was her last voyage to the Southern Ocean and the rest of her career was spent working in the northern hemisphere. The manning of *DII* was undertaken by staff from the Royal Fleet Auxiliary Service, seconded to work under NIO direction and paid for by NIO. Many of the officers and crew were surprised to find the ship going out into the ocean and then stopping around for the scientific work to be done and perhaps returning home without a foreign stop. In the early days, NIO finances did not allow *DII* to be maintained at sea throughout the year and most years she was chartered back to the Admiralty, largely for work by the Admiralty Research Laboratory and once to the Air Ministry to act as a weather ship. Even with this income she was forced to be laid up occasionally.

Through the close scientific relationship between Dr Deacon and Sir Edward Bullard at the Department of Geodesy and Geophysics at Cambridge, an arrangement was made that Cambridge should have the use of *DII* for research for two months every other year. The first of these cruises was in 1952 when Dr Maurice Hill examined the continental margins of the SW Approaches to the UK. Subsequently Hill ran five cruises in *DII*, many jointly with staff from NIO and other universities.

Cruise planning for the year ahead was an informal process held at NIO under the director when NIO staff made their bids for sea time. Sometimes these were short excursions to sea to test new instrumentation

Harry Moreton – Bosun.

Dick Burt – Netman.

and on other occasions were cruises into the Mediterranean or across the Atlantic. Usually each cruise was devoted to a single discipline although there were times when the disciplines were mixed. These longer cruises were often an NIO contribution to international projects such as the International Geophysical Year.

Initially she carried 15 officers, including 6 scientific staff, and a crew of 35. Her Bosun was Harry Moreton, who had sailed with *DII* since she was built and sailed on her successor RRS *Discovery* until the late 1960s. Weather-tanned and white haired, he had a fund of stories about the ship, and on one occasion in 1957, when it was difficult to work because of rolling in the swell he managed, from his memories of prewar, to rig the huge trysail with which *DII* was equipped. The sail was there so that she could get home if she ran out of fuel in the Antarctic! In later years he took great pride in running the ship's library, and also operated a laundry service for those scientists who could not operate the washing machines! Service like that is no longer a feature of seagoing. Initially *DII* carried a scientific officer who was on board for all cruises, but later his role was taken over by a Netman. Dick Burt, a seaman on board from before the 1950-51 cruise, became the Netman and oversaw deck operations until retiring in 1970. His contributions to helping oceanographers were immense and he devised numerous practical solutions for deploying and recovering gear that were later recorded in an internal IOS report. Both Harry Moreton and Dick Burt were awarded British Empire Medal (BEM) for services to oceanography in 1969 and 1970 respectively.

Life on board *DII* in those days was somewhat different from cruises today. There were a maximum of six scientific staff on board at a time. Dress code insisted that all officers and scientists had to wear a jacket and tie in the wardroom and for meals. In the earlier days a Sunday service was held in the sickbay, conducted by Henry Herdman, but this soon stopped.

DII was equipped with a heavy, steam-driven, deep-sea winch holding some 5,000 fathoms (9,150 m) of tapered steel wire, that could be deployed to the ocean floor for deep trawl work and for dredging, giving scientists the chance to investigate the ocean deeps in detail. On the foredeck she had two hydrographic winches bearing 3,500 fathoms (6,405 m) of 4 mm diameter wire from which water-bottle stations could be operated using the NIO reversing bottle, smaller nets could be deployed or a deep-sea underwater camera for photography of the ocean floor. There was also a smaller winch in the welldeck. For most deep-water operations it was necessary to keep the ship head to wind and stationary in the water. This was helped by the high and flat superstructure that enabled the wind from ahead to be balanced against slow revs from the main engine. There were, of course, no side thrusters available to help keep station, which involved considerable skill on the bridge.

Inboard, the main laboratory, originally for biological work, had a swing table, essential for preventing everything sliding off as she rolled, and all equipment had to be securely fastened down. Alongside was the chemical laboratory which had racks for water bottles where sampled could be taken and the temperatures read from the reversing thermometers. Another laboratory aft, the rough lab, was used for the sorting of the catches from the various nets and trawls. Here, in an atmosphere of formalin, the specimens were preserved for later examination ashore. In addition there were a geophysical laboratory, a photographic store and a darkroom.

From the start of her use by NIO, *DII* was equipped with a Kelvin Hughes 26J Deep Sea Echosounder, transmitting through the bottom plates of the ship. The accuracy of the depth was determined by a governor-controlled rotating arm which recorded onto a roll of damp iodised paper. This had distinct disadvantages in that the governor was not well controlled and the watch keeper had to keep a record of the number of rotations per minute, and that when the depth changed rapidly, the echo did not record on the paper. Following a lead from the USA, a Precision Echosounder (PES) was developed at NIO, using as a base a Mufax facsimile recorder designed for the reception of weather charts, in which the sweep speed was crystal-controlled, the display was linear rather than curved and an echo was always recorded on the paper. To avoid the blanketing of transmissions and reception by bubbles under the ship, the transducers were housed in a streamlined fish that was towed alongside below the surface. The PES was used on all cruises from about 1960.

A major problem in the early days of *DII* cruises was that of navigation. For some scientific studies it did not matter too much how accurately the position of the observations was known, but for others it was crucial. Around the coastline various radio aid navigation systems were available, but out in the central ocean, and in particular beyond the north Atlantic, normal navigation depended on morning and evening stars and noon sights of the sun. Of course these all depended on the absence of clouds so that often one could steam for days without an observation of position and have to run on dead reckoning. It was not until 1956 that a gyro compass was installed!

For the detailed navigation required by ocean current measurement and by geophysical and geological work on the ocean floor, it was necessary to lay floating marker buoys moored in depths up to 5,000 m. There always had to be a compromise between the size, and hence weight, of the mooring cable and the size of the buoy to support it. This led to our using piano wire for the cable and Dhan buoys carrying a pole with a flag and radar reflector (later enhanced by active radar transponders to give increased range). Unfortunately the relatively weak piano wire frequently broke if the currents were too strong or the seas were too rough. In any event, changeable currents could swing the buoy around, so that for the

Peter Foxton, John Swallow and Dick Burt laying Dhan buoy wire on Discovery II *in October 1955.*

very precise position measurements required by tracking deep-sea floats, movements of the buoys had to be established by repeated surveys of some identifiable feature of the bottom topography using echosounding.

For geophysical surveying, a network of buoys was sometimes laid with survey lines run out from the radar range of the buoys relying on dead reckoning before returning to radar range. Thus many of the surveys ended up as stars or boxes linked to the moored buoys. The scientists on board had to spend an appreciable part of their time struggling with these navigational problems before beginning to interpret the data collected. It was only later that radio (Decca Navigator and LORAN) and eventually satellite navigation systems were used.

Underwater acoustics played a large part in the data collection. Echo-sounding and sonar, acoustic telemetry, observation of deep sea floats, all required a minimum of background noise. We were very fortunate that *DII* was powered by a slow-moving reciprocating steam engine and so was very quiet. Indeed newer research ships in the USA and elsewhere used steam for the same reason. The increase in background water-borne noise in her successor, the diesel-electric ship, RRS *Discovery*, was very noticeable and caused some problems.

In January 1959, the National Oceanographic Council realised that the time was fast approaching when it would no longer be economical to run *DII* and that it was essential to replace her, and authorised the preparations of designs for a new ship. *DII* was programmed through to July 1962, when she was finally paid off. A good summary of her working life was written by Henry Herdman and published in the NIO Annual Report of the Council for 1962-63.

The detailed design for the new ship was masterminded by Herdman but with a substantial input from seagoing oceanographers at NIO. She had to be larger than her predecessor to take increased numbers of scientists and increasingly complex equipment, and to work in the tropics she had to have air conditioning (lacking in *DII*). Her design included many more laboratories, and facilities for experimentation such as a removable plate in the bottom of a well forward of the bridge, onto which acoustic instruments could be mounted. She was powered by diesel-electric generators that supplied not only the main propeller but also a bow thruster and the generator supplying the ship's services. The bow thruster was an innovation in shipping that had only been installed before in the RMS *Queen Mary*. This gave the *Discovery* good manoeuvrability and made station keeping easier. Heavy A-frames on the fore and poop decks and hydraulic cranes fore and aft made it possible to handle heavy loads over the side.

The name given to her by Viscountess Hailsham when she was launched in July 1962 was RRS *Discovery*. Many have since asked why not *Discovery III* ? By now the original RRS *Discovery* of Scott fame lay

RRS Discovery *in her original white colour in Mombasa in 1963 during the IIOE.*

alongside the Embankment in London and had been acquired by the Navy and renamed HMS *Discovery*. So in keeping with naval tradition the name RRS *Discovery* was once more available.

Built by Hall Russell in Aberdeen, RRS *Discovery* was 261 ft in length, with a beam of 46 ft and gross tonnage of 2,665. She had a range of 15,000 miles at an economic cruising speed of 10-11 knots. The facilities for science on board were considerably greater than those on *DII*. In addition to her twenty five officers and crew, she could take twenty scientists, who had access to a number of scientific spaces; a scientific plotting office with large plotting table, an oceanographic laboratory for water sampling, a biological laboratory, electronics laboratory, chemistry laboratory, scientific workshop, after rough laboratory, photographic laboratory and a gravimeter room. On deck were two large electric winches carrying 10,000 m of tapered wire, for coring and trawling, and three smaller steam-driven winches for vertical work, one of which had a conductor-cored cable. Initially she was painted all white, but following the Indian Ocean cruises her hull was painted black and her funnel yellow, the colours of the NERC ships today.

Following her pre-commissioning trials, she sailed to the Pool of London where she was on view to members of the National Oceanographic Council, Parliament, the Development Commission, the Board of the Admiralty and the public. Sailing upstream under the Tower Bridge, the Captain spun the ship in her own length before mooring using the bow thruster, much to the surprise and admiration of the Pilot.

Returning to Plymouth from London, a series of mishaps delayed the start of the trials cruise. At anchor in Plymouth Sound the ship's motor boat was being recovered when davit controls failed, and the boat slid down the side into the sea, bending the davits. There were also engine troubles requiring spare parts, so a diversion to Liverpool was necessary.

Peter David, Tony Laughton and Arthur Baker in the Plotting Room on Discovery.

A short trials cruise took *Discovery* down to the Salvage Islands, some hundred miles north of Tenerife, the crest of which could be seen over the horizon. We anchored in shallow crystal-clear water where in the sunny climate of this latitude all the scientific gear could be tested and fine-tuned. It turned out that there were problems with the sideways thrust on the flanges of the main winches which badly distorted when winding the wire in. Many other defects in the ship showed themselves and it was clear that *Discovery* would have to spend some weeks in refit, and undertake another small shakedown cruise before she would be fit to depart for the Indian Ocean.

In spite of all the initial problems, *Discovery* proved to be an extremely good ship from which to carry out oceanographic measurements. Her first programme was the International Indian Ocean Expedition in 1963-64 when she made three major cruises investigating many aspects of the NW Indian Ocean circulation, biology, geology and geophysics that are described in various other chapters in this book.

NIO had decided to base its ships at Plymouth, partly because this was a compromise between easy access from Wormley and proximity to the open Atlantic, which was our main sphere of operation, and partly because, being a fishing port, crew were available who had experience of operating gear over the side of a ship. When NERC was formed in 1965 it was decided to form a Research Vessel Unit to be responsible for all its ships, and this was initially also based at Plymouth with Henry Herdman in charge, assisted by John Cleverley. Staff were recruited to be responsible for the maintenance of seagoing equipment and

Photograph taken by Peter David of RRS Discovery *ship's company in Aden in August 1963.* ***1.*** *Tom Humphrey. Chief Engineer;* ***2.*** *Cdr. Bob Nesbitt, UK Hydrographic Office;* ***3.*** *Brian Irwin, Marine Laboratory, Aberdeen;* ***4.*** *F.A.J. Armstrong, Marine Biological Association, Plymouth;* ***5.*** *Harry Moreton, Bosun;* ***6.*** *Ron Currie, NIO;* ***7.*** *Don McKay, Assistant Electrical Officer;* ***8.*** *Tom Tucker, NIO;* ***9.*** *Bobby Forsyth, Junior Engineer;* ***10.*** *Andy Kaye, Junior Engineer;* ***11.*** *John Swallow, NIO;* ***12.*** *Ken Douglas, 2nd Engineer;* ***13.*** *Brian Briggs (Assistant Bosun?);* ***14.*** *Peter David, NIO;* ***15.*** *Peter Herring, Cambridge University;* ***16.*** *Betty Kirtley, NIO;* ***17.*** *Tim Vertue, NIO;* ***18.*** *Arthur de C Baker, NIO;* ***19.*** *Andrew Bryce, Electrical Officer;* ***20.*** *John Scott, 2nd Officer (Grandson of R.F. Scott);* ***21.*** *Roger Bailey, Ornithologist, Oxford University;* ***22.*** *Capt. C. Alexander, D.S.C., Master;* ***23.*** *Tony Chapman;* ***24.*** *??;* ***25.*** *Graham Topping, Liverpool University;* ***26.*** *Martin Angel, Bristol University;* ***27.*** *A. Prakash, Biological Station, St Andrews, Canada;* ***28.*** *Peter Brewer, Liverpool University;* ***29.*** *Tony Green, Junior Engineer;* ***30.*** *Alec Redpath (4th Engineer);* ***31.*** *Mike Bretherton (3rd Engineer);* ***32.*** *??*
Not included*: R.M. Frederick, Chief Officer; Eric Anstead, 3rd Officer; Tony Champion, Radio Officer; Malcolm Kelly, Ship's Doctor; Harry Kelsey, Chief Steward; D.H.Cushing, Fisheries Laboratory, Lowestoft; A.D. Macintyre, Marine Laboratory Aberdeen; A.F. Boxell, NIO; N.D. Smith, NIO; A.R. Stubbs, NIO.*

instrumentation. The Unit moved to Barry in South Wales in 1969, and after a year or two, a new building was constructed and opened in 1972. The following year, when IOS was formed, the Marine Scientific Equipment Service there became a separate entity with Dr L.M. Skinner in charge and responsible to Tom Tucker, by now the Assistant Director, IOS Taunton.

Up until 1965-66 Henry Herdman was in charge of the research ship operations of NIO, but after a driving accident he had to retire and his role

Henry Herdman (Left) and Arthur Fisher (Right)—two key people in the operation of NIO's research ships.

was taken over by Arthur Fisher who had been working with the biological group since 1953, but who rapidly established himself in his role as officer in charge of the research ships. On the formation of NERC, ownership of *Discovery* passed to NERC Council, although management responsibility remained with NIO. Eventually programming and management passed to the Research Vessel Service of NERC.

Discovery had a major refit at her builders from October 1968 to January 1969. The years since the IIOE, with the introduction of profiling conductivity-temperature-depth (CTD) instruments, the need for more sophisticated navigation and the general need to record data, had given us a clear idea of the future capabilities we needed in the ship. A major change in the layout of the lab space was needed and this included the requirement for a dedicated computer room. Computers then were rather large. They required air conditioning and also extensive provision for present and future data wiring between lab spaces and from permanently installed logging instruments.

The computer installation was overseen by Jim Crease during the refit with great co-operation from the Hall Russell workforce both individually and collectively. For example, when they 'knocked off' for the day after fitting the wiring conduits in the new labs they would leave their personal tool-boxes open and accessible so that the staff from NIO could work overnight on the scientific inter-wiring. This co-operation was in no small measure due to Arthur Fisher who had by then taken over a role that was essentially that of 'ship's husband' whenever we came into port, a position he retained for many years. He established such a rapport with management and staff that he was a guest of honour at Hall Russell's Burns' night supper in January 1969.

The choice of the shipboard computer hinged on the machine's robustness and likely ability to tolerate the pitch and heave of a small ship. It also had to accept and integrate analogue and digital signals from a variety of sensors. The eventual choice of the IBM1800 (an industrial process computer) turned out to be a good one. Our colleagues at the Scripps Institution of Oceanography were also following this route. The Scripps machines used magnetic tapes for storage whilst we used 3 removable disk drives, each with the then-large capacity of 1 Mbyte!

With a computer on board it also became possible to exploit the developing satellite navigation systems. Civilian equipment for the US Navy's Transit satellite system was just becoming available and NIO acquired the 10th production model of a Magnavox system. Programmes to acquire the Doppler-shifted data stream and ephemeris (orbital parameters and timing information) from the satellite were available but we had to write all the analysis software. We felt we were ready to take the system to sea when we fixed the position of the Wormley lab to within a few feet of that given by the Ordnance Survey National Grid. Individual fixes at sea gave an accuracy of about 200 m.

Fitting the equipment onto the ship was not without incidents. Jim Crease recalls that when the new computer was being pushed on a temporary steel platform from the dock through a hole in the ship's side the tack welding of the platform started to give way! IBM insured the equipment whilst it was ashore and we were responsible once it was on the ship. We were facing a nightmare scenario until a final push had the machine safely on board. Even the apparently simple task of installing the satellite navigation antenna on the foremast provided drama. Jim, a fitter and the antenna were in a bucket suspended from a shoreside crane when the ship started to move back and forth as a seiche developed in Aberdeen harbour. Jim hung on and the antenna was fitted!

The only aspect that IBM would not guarantee was the behaviour of the disks as the ship pitched and rolled and slammed. In fact the drives demonstrated their robustness on the first cruise after the computer was installed. In a particularly vicious storm a wave hit the ship, carrying away a davit on the foredeck, bending the bulwarks and shattering portholes: the disks continued to operate perfectly.

We installed an identical IBM machine at Wormley to continue developing the software while the ship was at sea, testing instrument interfaces, and developing an in-house data processing capability. This led eventually to establishment of the national Marine Information Advisory Service under Meirion Jones and the Marine Sciences computing facility at the

Proudman Lab under Brian Hinde, who later became Director of NERC Scientific Services after joining us as a Scientific Assistant.

The frequency of satellite fixes was every hour or two dependent on satellite location and the number in operation. The fixes, unlike the present GPS system, depended on using the ephemeris of the satellite and the Doppler shift of a very stable frequency it broadcast as it passed overhead and abeam. Waiting for the satellite became a common feature of life at sea. The ship's track between satellite fixes was interpolated by using the two-component E-M log and compass to provide 'dead reckoning (DR)'. The closure error between the end of a DR track and a new fix was ascribed to the surface current between fixes, although this 'current' also compounded various measurement errors which over succeeding years became better understood.

This development ushered in an era when knowledge of the ship's position was no longer an issue in our experiments. In the early days, Geoff Howe, Master of *Discovery* and a great supporter of our work, put up a very good defence of the traditional method of position fixing by star sights (although only at dawn and dusk!). Increasingly over the years, we became able to use combinations of satellite receivers to better compute the heading, ship's speed over the ground and position at any time and any place.

Over time, numerous other instruments such as the magnetometer and a suite of meteorological sensors were added as routinely logged variables, although we quickly learned that automatic data logging in no way dispensed with the need for continual care and maintenance of the sensors.

Going to sea and living in close proximity with friends and colleagues for weeks or months at a time played a huge part in developing the happy family atmosphere that was such a feature of life at NIO. Whilst the chief scientist and the Master were always 'in charge' there was essentially no distinction on board between the senior scientists and the most junior technicians, and similar rapport was built up between the scientists and the officers and crew, many of whom served on board for many years.

Life at sea was very much team work. Work was continuous around the clock, with shifts according to the requirements and preference of the principal scientist. Biologists tended to do 12-hour day and night watches changing over at port calls so that everyone experienced at least some daylight! Teams of people got very good at deploying, recovering, repairing and operating the plethora of sea-going equipment in all sorts of weathers. Protective clothing and safety rules were non-existent or minimal, but very little equipment was lost and there were no serious injuries to people. Luck perhaps, but experience built up over years played its part.

When not working, people found diverse ways to entertain themselves and others. The social hub of the ship was the bar and mess room where 'talking shop' was frowned on. Here people unwound from the rigours of the deck, often playing assorted games: liar dice was very popular with the loser providing the drinks. Music and singing featured: there was a *Discovery* song book and some extremely good musicians with assorted guitars, banjos and squeeze boxes. This was before the days of music centres. On one occasion the old piano was laboriously carried up two decks from its home in the ship's library to the bar, but the resultant addition to the band was not worth the huge effort involved! Celebrations for anything were held and end-of-cruise parties were hosted by the principal scientist, who traditionally issued an RPC ('Requests the Pleasure of your Company') invitation to all.

Film nights were a welcome diversion. Films were 16 mm on large spools that had to be manually changed about twice during a show, providing endless opportunity for mistakes and unhelpful advice. Fancy dress parties, darts and table tennis tournaments with the crew, the captain's quiz nights (the first question was always "What is the opening line of Moby Dick?"), deck barbeques in the tropics, one of which featured a birthday cake consisting of a beautifully iced and disguised condom that exploded on being cut; all were part of life at sea. And there were always the opportunities for simply observing the oceans and the fascinating animals that swam or floated by. Whale and bird logs were kept; petrels

The Discovery *band on a biology cruise in 1969. (Left to right)*
Back row: *Greg Phillips, Mike Thurston.*
Standing: *Roy Wild, Bob Morris, Colin Hayes, Adam Locket, Peter Herring, Nigel Merrett, Howard Roe, Mike Longbottom, Malcolm Clarke.*
Sitting: *Julian Badcock, Bob Spencer, Philip James, Arthur Baker, Bob Aldred.*

and shearwaters tended to fly aboard at night, attracted by the ship's lights, and were gathered up into boxes and released in the morning. Flying fish were not so lucky: unpeeling dried corpses from the scuppers was common. Shark and squid fishing was entertaining, and luckily in the case of some very large sharks usually unsuccessful!

Port calls about every three weeks provided welcome diversion and mail from home, which was dished out in a large pile in the plotting office. Ports were sometimes mundane, sometimes exotic, and the early cruises were before the advent of mass tourism. Hiring taxis to go round islands like Madeira, the Azores and the Canaries provided great diversion. On one occasion a Senegalese taxi driver was paid with Monopoly money, which he indignantly returned to the ship the following day, and bargaining for African masks, beads and parrots using the ship's soap as currency was a feature of the quay in Dakar.

Although RRS *Discovery II* and *Discovery* were the research ships most used by NIO, scientists very often took part in cruises of other research ships, sometimes chartering for specific purposes and at other times working cooperatively with groups from other laboratories, especially in the international programmes. Getting appropriate international permissions and getting equipment and people to ports around the world often presented major logistic problems, not least in dealing with foreign customs and port officials. Much of this was master-minded by Arthur Fisher. Advanced planning was essential to avoid costly delays but the support which the research teams had from the logistics side of NIO meant that nearly all our objectives were met.

18

The library – a key research tool

Pauline Simpson

A well-organised library through which researchers can easily access both peer-reviewed and less accessible literature is an essential resource for any world-class laboratory and this was certainly true in the era before journals were accessible via the World Wide Web. NIO scientists were blessed with a world-class library run by librarians who were sympathetic to their needs and tolerant of their many demands and foibles.

The need for a library was specified in the Royal Charter incorporating the National Oceanographic Council, in 1950. Among the objects for which the Council was established was ... (d) "*to collect, collate and publish information and discoveries in all branches of oceanography and connected subjects*" and (e) "*to collect and maintain libraries or museums containing books, papers or objects connected with oceanography and all matters connected therein*".

The collection

Such a library was established in 1953 when the National Institute of Oceanography moved to Wormley, building on a number of pre-existing collections. The physicists from Group W/NIO at Teddington brought a large assortment of reprints and reports devoted mainly to physical oceanography. The biologists brought with them the Library of the Discovery Committee which contained many books on marine biology and, most importantly, a very valuable collection of reports on the scientific results of oceanographic expeditions. There were also publications that had been acquired by the Admiralty Centre for Scientific Information and Liaison (ACSIL) for the NIO following its formation in 1949. Added to this material were the extensive papers and reports on regional and physical oceanography collected personally by Dr J.N. Carruthers,[1] before and during his service in the Oceanographic Branch of the Hydrographic Department. There were other important additions during the early 50s:

Key people in the NIO Library. ***Left:*** *J.N. Carruthers.* ***Right:*** *Mary Swallow.*
Facing page: *D.W. 'Dick' Privett.*

the D.J. Matthews Collection was purchased, comprising books and reprints, and the Library also became the custodian of the Challenger Society card catalogue of the '*Bibliography of the Marine Fauna*' which lists works published between 1758 and 1907.

These acquisitions, and particularly those from the Discovery Investigations and from Dr Carruthers, meant that the library could claim comprehensive coverage from the late 19th century. A definitive collection continued to grow thanks to the budgetary support from NIO management. The Discovery Committee Library also included many early oceanographic texts and a large collection of Expedition Reports. These included not only original copies of the Challenger Expedition Reports but also the reports of many national expeditions carried out by ships, the names of which (*Belgica, Investigator, Siboga, Meteor, Percy Sladen, Erebus and Terror* etc.) conjure up the spirit of scientific exploration in the late 19th century.

The Discovery Committee also contributed their monograph series *Discovery Reports;* the principal results of investigations in the Antarctic and other waters in the southern hemisphere undertaken by the Discovery Committee. With these came the Discovery Committee's publication exchange agreements with similar research institutes worldwide. The *Discovery Reports* series (the last was published in 1980) provided NIO Library with a ready-made exchange 'revenue'.

The contributions from Dr Carruthers warrant further mention. He treasured books and collected thousands of reprints, serial publications, books, atlases and charts from his colleagues worldwide, before, after and during his work in Germany after WWII, when he was advising and supporting the German oceanographic community in protecting its

infrastructure and libraries from acquiring nations. So important was his contribution to the rapid rebuilding of German marine science that he received many honours from Germany.[1]

NIO also inherited Dr Carruthers' card catalogue; a detailed index not only to his own collection but of the wider literature as well (he was in the habit of cutting up contents pages and sticking them onto cards). Whenever enquiries for data or information came to NIO, he was able to answer from his own wealth of knowledge backed up by references from his catalogues. He tried to be as definitive as possible, primarily to enable him to answer enquiries with published source data, but also out of pure love of the task – his indexing terms were uncontrolled – 'Men swallowed by fish (a pearl diver swallowed by a giant bass)'; 'Corpses afloat: Biscay, Rennell's Current'; 'Catholic faith and fish consumption'. Through the years the Library has been able to answer some very strange enquiries only because of a Dr Carruthers-era index entry!

Each year the NIO Library bound together volumes containing copies of all papers published by its scientists. These were known as the Collected Reprints. All senior scientists received copies and they were also exchanged for equivalent volumes from other laboratories. The first, prepared in December 1953, contained the 61 papers published since 1949. The series continued for 21 years.

The Librarians

As NIO Information Officer with an office in the Library, Dr Carruthers played an important part in suggesting what would now be termed a 'user-friendly Library' – a simple system of 'classification' and locations. He continued contributing to the indexing of marine science literature right up until the day he retired.

NIO's first Librarian from 1953 was Mary Morgan (who later married John Swallow).[2] Her lasting contribution was to greatly extend the Discovery Committee's exchange programme. Using *Discovery Reports, NIO Annual Reports* and an *NIO Report Series* as 'revenue' she built up a worldwide publication exchange network with some 900 organizations which continued until the World Wide Web replaced the mailing of publications, by institutions uploading their publications onto their websites. From 1963 to 1977 she was (jointly with Dr Mary Sears at Woods Hole) the Editor of *Deep-Sea Research*, at that time the foremost marine science journal. Having an 'in-house' editor was a great help to NIO's scientists.

In 1959 Dacre (Dick) Privett took over as Librarian. Dick had been a member of the original Group W at ARL, researching air-sea interaction. He brought order to the Reports Collection of over 100,000 items and initiated control on the cataloguing and indexing procedures. His interest in indexing finally culminated in his being the main compiler of the Aquatic Sciences and Fisheries Information System (ASFIS) Thesaurus published in 1985 which is a controlled publication indexing vocabulary for marine science still used worldwide.

The Library staff grew and in 1969 the NIO's first professional librarian was appointed as Assistant Librarian followed by an Information Scientist, hired to support the indexing of books, monographs, journal articles, conference papers, etc. adding to the unique retrieval opportunities not offered by any other marine science library in the world. This detailed catalogue predates any commercial marine science database (the earliest indexed paper is 70 AD!), underpinned the Enquiry Service and eventually supported the Marine Information and Advisory Service.

The library's role in NIO and beyond

Both in its initial location on the second floor of the main building and, from 1966, sharing with the lecture theatre the top floor of the new block (ironically without a lift), the Library was a hub of activity in a way that has changed greatly since automation and the ready access to online journals. The Library and its staff provided access to the latest journals, inter-library loans, charts and atlases, a publications service as well as to historical documents and of course the enormous card catalogue and reprint collection. It was also a quiet refuge to study – or to read the newspaper.

The NIO Library gave access to a much wider information network of the shared resources of other libraries. Initially this had a naval focus firstly through the RNSS and eventually based around the Naval Science and Technology Information Centre (NSTIC) services. When NIO came under NERC in 1966 a grouping was built from libraries in NERC laboratories and other governmental departments with an interest in marine science, e.g. MAFF, MBA and the Fisheries Laboratory Aberdeen. The UK Marine Science Librarians Group was formed in 1969 with NIO Library as a founder member and the Library took part in a number of cooperative projects: union list of serial holdings, a directory of members libraries, etc. Membership of national (BIASLIC) and international (IAMSLIC) marine science information networks continues to this day.

The library in the new building (after 1966). Part of the extensive card catalogue can be seen.

The present-day National Oceanographic Library in Southampton is a direct continuation and expansion of the Wormley Library. Its archives include the papers of the Discovery Committee, the National Oceanographic Council, past Directors of NIO, SOC and of the Challenger Society for Marine Science as well as a rich and varied photographic record. Today's researchers, students and historians alike still reap the benefit of the Library founded at Wormley.

19

The 'backroom boys'

Anthony Laughton

Administrative support

Behind every inspiring and successful director, there is an administrative structure that enables him to carry out his vision. Dr Deacon was very lucky to have such support in the person of R.G. Williams, who was Secretary of NIO from 1955, succeeding Capt. R.H.G. Franklin and W.J. Hanman. Ray Williams oversaw the development of NIO, the financing of the new ship, the transition to NERC funding, and the change of name to IOS, involving the incorporation of the Institute of Oceanography and Tides (ICOT), the Unit of Coastal Sedimentation (UCS) and the Marine Scientific Equipment Service, finally retiring in 1983.

The role of an Institute Secretary was far ranging, but under Dr Deacon was based on two simple principles: first, everything had to be geared to the overriding importance of facilitating the scientific work, and second, the scientists should be protected from bureaucratic interference. Predominantly, he was responsible for the funding and expenditure of NIO, accountable through the Director initially to the National Oceanographic Council (NOC), and later to NERC. But he was also responsible for staff recruitment, pay and conditions, for the operation of the office, for stores and for the buildings and their maintenance.

The NOC had initially appointed Captain Franklin as Secretary but he was succeeded shortly after by Bill Hanman on secondment from the Admiralty. But when Hanman was recalled to the Admiralty, the NOC realised the importance of the continuity of a permanent appointment, so the post was again advertised and Williams was appointed in 1955. Hanman was responsible for the complex negotiations for the acquisition of the Wormley site and the setting up of systems and procedures of the fledgling NIO.

Williams had read history at Cambridge, and had initially planned to go into teaching, but later became a temporary civil servant in the finance division of the Admiralty. His role at NIO was widely appreciated by the members of staff. He was always approachable and guided newcomers and old timers through the intricacies of how to go about purchasing equipment, how to arrange travel, how to square their expenditure with their budget, how to go about getting promoted, how to get typing and photography done and the thousand and one things that make for efficient running of the laboratory.

Finance and staffing

The main source of income for NIO was initially a grant-in-aid of £110,000 (equivalent to £2.8 million in 2010 money) under vote 6P of the Navy estimates. This comprised £50,000 from the Admiralty, £50,000 from the Development Commission, and £10,000 from Colonial and Middle Eastern Services. Small grants were also received from Australia, New Zealand and Ceylon.

The scientific and experimental staff numbers in 1949-50 were 29, with five administrative staff, excluding the 16 officers and the crew who manned the research ships *Discovery II* and *William Scoresby*. The staff were members of the Royal Naval Scientific Service, in line with those who originally worked at ARL, and seconded to NIO. Those in Scientific Officer grades were expected to initiate research, whereas the Experimental Officer and Scientific Assistant grades were there to support it. Later, under NERC, these distinctions were removed and all were Scientific Officers of various grades. Although this was a move towards equality of status, in fact it meant that the former Experimental Officers had to be judged for promotion in competition with the Scientific Officers.

The annual income hardly increased in the years up to 1955, staying in the region of £120,000 per year, although by that time the scientific and experimental staff had increased to 50 and the administrative staff to 14 (a number that included direct support staff such as secretaries, typists, storemen etc.). Staff salaries and costs in general were rising.

Ray Williams, NIO Secretary, photographed at his retirement.

The costs of moving into the buildings in Wormley were significant. So it was not surprising that it was very difficult for the Secretary to make ends meet and it was necessary to lay up *Discovery*

Staff and expenditure at NIO 1949–1973 (Based on NIO and NERC Annual Reports)					
Year	Scientific staff	Admin. Staff	Industrial staff	Annual expenditure (£k)	Comments
1949-50	29	5	Not listed	49	
1950-51	32	5	Not listed	125	
1951-52	35	5	Not listed	124	
1952-53	34	8	Not listed	138	
1953-54	44	9	Not listed	109	
1954-55	49	9	Not listed	146	
1955-56	52	10	Not listed	170	
1956-57	50	13	Not listed	193	
1957-58	53	12	Not listed	201	
1958-59	53	14	Not listed	222	
1959-60	60	15	Not listed	240	
1960-61	69	17	Not listed	283	
1961-62	67	19	28	637	*Discovery* build
1962-63	72	18	27	760	
1963-64	78	24	29	401	
1964-65	87	25	30	442	
1965-66	97	26	29	490	
1966-67	Not listed	Not listed	Not listed	654	
1967-68	115	27	37	737	
1968-69	Not listed	Not listed	Not listed	928	
1969-70	Not listed	Not listed	Not listed	867	
1970-71	Not listed	Not listed	Not listed	953	
1971-72	Not listed	Not listed	Not listed	898	
1972-73	Not listed	Not listed	Not listed	1,108	

II for twelve months in late 1952 after her Antarctic cruise and, until more funds became available, to operate her for NIO science only six months in the year, the rest of the year being under charter to the Admiralty, and once to the Air Ministry to relieve the weather ship at station 'Kilo'. After her whale-marking work in 1950, it was not possible to operate *William Scoresby* and she was laid up until she was sold in 1954.

It was not until 1957 that the income from the Admiralty and the Development Commission was raised to £175,000 for the following five years. Another increase from these departments of £25,000 in 1961, together with smaller grants from Commonwealth countries and from other Government Departments for special research projects, repayment work and charter out of *Discovery II*, brought the total income to nearly £300,000.

However, during the following four years special grants were made to NIO to build a new research ship to replace *Discovery II*. The new ship, *Discovery*, was launched in July 1962. Construction by Hall Russell in Aberdeen was overseen entirely by NIO staff. Herdman looked after the construction, and Williams monitored the expenditure.

The total cost was just short of £900,000 (£7 million in 2010 money). The design was so successful and the cost so low, that the Hydrographic Department of the Admiralty built three survey ships, HMS *Hydra*, *Hecla* and *Hecate* to roughly the same design for the price of the one naval replacement for HMS *Owen*.

Williams remembers the occasion when he accompanied Deacon, Charnock and Herdman for a meeting, supported by detailed documentation, with Admiralty and Treasury officials (referred to by Herdman as "*three black crows*") to make the case for a grant for a new research ship. The senior official opened by asking "*Why does an Institute of Oceanography need a ship?*" The NIO members could never be sure whether he had read the papers or was just trying to throw them off balance. But the right decision was made in the end.

By 1962, the total staff numbers had increased to about 90, over twice the numbers of 1952. This had implications for accommodation and for the facilities that were needed for research.

Facilities at Wormley

When NIO was first created in 1949 its staff were distributed between six locations:

Admiralty, Queen Anne's Mansions, London
Queen Anne's Chambers. London
Natural History Museum, London
Admiralty, Hydrographic Department, Cricklewood
Admiralty Research Laboratory, Teddington
Marine Biological Association, Plymouth

The move to Wormley, Surrey, in February 1953, brought together the far-flung staff for the first time and provided the focus for the growth of NIO during the following decades. The four-storey building had been built for the Admiralty Signal Establishment during the war, especially to develop and test naval radar. On the roof was a rail track for manoeuvring radar sets and inside was an enormous lift, accessed by roller doors three stories high. Surrounding the building were various black Nissen huts which were used for storage of equipment and the supply of stores under the eagle eye of Mr Baxter. Naturally many alterations were needed at the start to create workshops, a library and open-plan laboratories for physics and for biology.

In due course, and when finances permitted, many further alterations were made to the buildings. The enormous rolling doors were removed (although the large lift was retained throughout), the rails removed from the roof and a fifth floor was added. Later on the Black Hut (the Nissen Hut used for storage) was removed and replaced with an entirely new three-storey building. In the basement area was a wave tank to study waves and their effect on sediment transport, a pressure facility for testing housings to go to oceanic depths, acoustic tanks, new workshops and drawing office. A new administrative area, overseen by Miss Stiven, was located on the first floor. On the top floor the expanded library, the domain of Dick Privett, and a conference room provided essential support for the science. The new building was opened in September 1966.

Critically important groups in support of the science were those of the Drawing Office and Engineering Workshop. Very little instrumentation for oceanography could be bought commercially, so new instruments, initiated in the science groups, had to be made in house or by contract. Technical drawings were essential and much of the manufacture was done in the NIO workshops. NIO was very fortunate in having Frank Pierce to head up the Engineering Group, which worked closely with the Applied Physics Group. Pierce had been a design engineer at the Admiralty Gunnery Establishment, and was in Group W before the formation of NIO.

By 1968, the last date for which a staff list is available, the numbers of scientific staff had increased to 115, administrative staff to 27 and industrial staff to 37 – a total of 179. Another 13 staff were employed as officers on *Discovery*.

The Natural Environment Research Council

On 1 June 1965, NERC was created following the Science and Technology Act, 1965. NERC was to become responsible for overall policy and finance of the entire field of marine research in the UK. The NOC petitioned the Queen to accept the surrender of the Royal Charter, and formally came to an end by virtue of Order in Council of 31 January 1966. Under the Oceanography and Fisheries Committee of NERC, a special committee was set up to look into the affairs of NIO. Financial control was removed from the Director and Secretary of NIO and transferred to NERC Headquarters. This was a blow to the independence of NIO and particularly to the Director. NERC had to demonstrate that it was more capable of generating original research than NIO had been as an independent body.

In due course, the staff at NIO were transferred from the Royal Naval Scientific Service to membership of NERC where they were Public Servants (as contrasted to Civil Servants, but enjoying the same conditions of service and pay).

IOS buildings in the 1980s. Over the entrance to the new (LH) block is the figurehead from HMS Challenger *(just visible).*

Under NERC, the usual NIO Annual reports ceased to be published, the last one covering the period 1966-67 and 1967-68. For the next five years the affairs of NIO were covered as part of the NERC Annual Report. The staff numbers grew from 179 to over 230 and expenditure also grew from £490,000 in 1964-65 to £1,100,000 in 1972-73.

Deacon retired as Director in 1971 and his role was taken over by Professor Charnock. By the time that NIO had been combined with other NERC laboratories on 1 June 1973 to form IOS, the total number of staff in all laboratories had increased to nearly 400, some 240 of which were at Wormley. The era of commissioned research, in which funds for NIO had to be earned by contracts with Government Departments and with industry, was just over the horizon. The bureaucracy and administration involved in running the programmes had increased enormously in line with accountability since the foundation of NIO. Although the funds were never as tight as in the early days, the freedom for frontline researchers to follow their instincts was diminished.

Working within NERC involved an additional bureaucratic burden but Williams was effective at protecting NIO's scientists from this. He skilfully led the NIO supporting services throughout these years of expansion and change until his retirement in 1983.

Beyond 1973

This book formally ends in 1973 with the linking up of NIO with the Institute of Coastal Oceanography and Tides at Bidston, the Unit of Coastal Sedimentation at Taunton and the Marine Scientific Equipment Service in Barry, and the subsequent change of name to the Institute of Oceanographic Sciences. However the work of NIO continued and expanded under its new name and the new administration under NERC. NIO left a legacy, inspired largely by the directorship of George Deacon, which continues today at the National Oceanography Centre at Southampton. Anthony Laughton and Howard Roe describe these changes and the legacy that NIO has left to our current understanding of the role and importance of the oceans to us all.

20

The legacy

Anthony Laughton and Howard Roe

Everything changed in 1965. Deacon, in his notes to the Royal Society, comments (in the third person) on the take-over of the operation of NIO by NERC:

> *Deacon was convinced that this would lead to over-centralized administration and that those doing the work would find it more and more difficult to have any say in policy making. There was more money but less direct responsibility and involvement and this was a set back for a laboratory that had gained an international reputation for doing great things at relatively little cost, and later on as detailed accounting, administration and overheads added to the labour and cost of even minor transactions, exchanges with other laboratories and even between related projects in the institute itself were hindered by formalities, and it became much more difficult to share expensive facilities and hard won experience.*

The change came about because of the Government's decision to create a number of research councils to administer scientific research in the UK, under the Science and Technology Act of 1965. The National Oceanographic Council, which had hitherto administered NIO, formally ceased to exist on 31 January 1966 after it had petitioned the Queen for the surrender of the Royal Charter. Deacon never forgot the formal defacing of the Royal Seal. The management of RRS *Discovery* was transferred in 1969 to the newly-created Research Vessel Unit located in Barry, South Wales.

The research programme continued, but with responsibility to the Secretary of NERC, Ray Beverton. Deacon had to work with NERC but he never felt particularly happy about this. He retired in 1971 with a knighthood, and the search started for a successor. One obvious candidate was John Swallow, who had already distinguished himself with his neutrally buoyant float and the measurement of deep-ocean currents. But he was unwilling to take on an administrative role since he wished to pursue his research interest in deep-ocean circulation.

The new Director was Henry Charnock, who had been part of the original wave group of NIO at ARL. However, he had left NIO in 1958 to take up a readership in his old department at Imperial College, only to resign from it a year later when he started a new research group for NATO at La Spezia in Italy, as a seconded staff member of NIO. He returned to NIO in 1962 to pursue his work on air-sea interaction. But he was not at ease there, having some conflicts with Deacon, and he took up the new chair in Physical Oceanography at Southampton University. There he expanded the field of oceanography at Southampton from the near-shore to the open ocean, and set up, in collaboration with NIO, some of the preliminary cruises for the JASIN experiment of 1978.

Cartwright wrote about Charnock in the Biographical Memoirs for the Royal Society:

> *The laboratory at Wormley had greatly expanded and was in need of internal reorganisation: this was Henry's first task. The NERC had wanted to disband the marine biology group, but, agreeing with Deacon on the necessity of a multi-disciplinary basis for marine science, Charnock insisted on keeping it at NIO.*

He had further bureaucratic demands made on him by NERC, which in the face of looming recession in 1973, merged NIO with three other smaller laboratories in the marine field: the Institute of Coastal Oceanography and Tides at Bidston, near Liverpool, the Unit of Coastal Sedimentation located alongside the Hydrographic Department in Taunton and the Marine Scientific Equipment Service (MSES) attached to the Research Vessel Unit in Barry. To manage the laboratories at Bidston and Taunton, Charnock appointed respectively David Cartwright and 'Tom' Tucker, both from NIO, as Assistant Directors. Responsibility for the MSES was given to the Assistant Director at Taunton. NIO and its new components were named the Institute of Oceanographic Sciences. It is at this juncture that this history formally finishes.

However, the legacy of the NIO and the international status that it had acquired under Deacon, and later Charnock, lived on in the greatly expanding scientific programmes of its successor laboratories under subsequent directors.

In 1972, the government had implemented the recommendations of the Rothschild Committee, that some of the publicly funded bodies should, where they were doing work for government departments, have funds transferred to those departments who would in turn provide funds for the contracted work. In fact, the Rothschild principle never worked as intended and resulted in a loss of funding to the Institute that departments were reluctant to offset. Initially, this principle applied only to transfers within government but it later developed into a customer-contractor relationship with private industry and overseas bodies as well. In order to maintain the flow of funds to run IOS, bids had to be made for contracts wherever they could be found and this became increasingly difficult. As a result, the scientific programmes became progressively divided between those done as 'blue skies' research and those done for customers as 'commissioned' research.

Charnock's management style was greatly different from Deacon's. He instituted groups covering each discipline and appointed group heads. Rather, however, than have a hierarchical structure, he operated what he called 'orbital management'. Group heads were able to operate at a distance from him, but occasionally their orbit would come close to the centre when he would exercise his influence.

Charnock spent a lot of his time arguing for new contractual work with potential customers. He was particularly adept at chairing meetings and coming to quick decisions. In the words of Tony Rees, who had been a close colleague at Southampton University and later worked at NERC headquarters on commissioned research before he joined IOS, he *"was able to argue his case, often before a potentially hostile audience, good humouredly and so cogently and in such reasonable sounding terms that he usually came out a winner.... He won a lot because he was usually right."*

Charnock won an important commission, which was to last for eight years, from the Department of the Environment. This was concerned with a feasibility study for the possible disposal of high-level radioactive waste in the deep ocean. An international programme under the Nuclear Energy Agency of the OECD, set out to discover whether cylinders of heat-generating vitrified waste could safely be disposed of tens of metres below the sediments of the deep oceans in regions of well-established stability. The pro gramme involved knowledge of the sediments themselves and any cracks or faults within them, of the role of deep currents dispersing and diluting any possible radioactivity escaping from the seabed, of the uptake of radionuclides by the biota in the ocean and finally the routes by which these could reach

Henry Charnock early in his career.

mankind through fisheries or marine organisms generally. The programme was well suited to the wide range of oceanographic disciplines within IOS and was a justification for maintaining them.

Another big contract in IOS was for Applied Wave Research. This was not a simple single contract and though mostly with the Department of Energy (DEn), some was with the Department of Industry. For most of the first ten years of IOS, about ten scientific and technical staff were employed on this contract mostly at Taunton in the Applied Wave Research Group headed by Ted Pitt, but also including David Carter, Peter Challenor and Laurie Draper at Wormley. Apart from advising the DEn on regulations for the offshore industry, these groups did a lot of work for the wave-energy programme.

In another field, the breadth of IOS expertise was used to advise the Foreign and Commonwealth Office in the protracted international negotiations with United Nations Conference on the Law of the Sea (UNCLOS). Over the years since then IOS and its successor organisations have played an increasing role in helping, not only the UK, but also foreign countries on their rights and responsibilities under the UN Convention on the Law of the Sea.

In 1978, with the prospect of having to retire at the age of 60, Charnock decided to leave IOS and return to his old chair at Southampton University where the retirement age was 65. He later became closely involved with a House of Lords Select Committee report on the state of marine science in the UK, which had significant consequences for IOS. After an interim period when Peter David acted as director, Anthony Laughton was appointed to lead IOS, a post which he held for the following ten years.

After several years of planning, the major international programme JASIN (Joint Air-Sea Interaction) took place west of Scotland in 1978, involving fourteen ships, three aircraft and nine countries, and coinciding with the launch of the first dedicated oceanographic satellite, SeaSat, for which JASIN provided a great deal of the ground truth. This was both a step in the increasing awareness of the importance of the oceans and their interaction with the atmosphere that was critical in assessing climate change and, through satellite observations, the start of an era of truly global-scale ocean observations. IOS and its successor laboratories later took part in the World Ocean Circulation Experiment (WOCE) and the Climate Variability and Predictability studies of the World Climate Research Programme (WCRP), and hosted their international project offices. All of this led up to the development of new models of ocean circulation and an understanding of the influence of the oceans on climate.

Current measurements for JASIN in the Rockall Trough attracted the interest of oil companies, who were looking to expand their operations from the North Sea to deeper and more environmentally challenging areas north and west of the UK. These resulted in continuing investigations by IOS of currents in deep water NW of Shetland supported by a consortium of oil companies and by the UK Department of Energy.

The early work on surveying the deep-ocean floor with GLORIA Mk I persuaded Dr Deacon to support the development of a much more user-friendly Mark II which was completed in 1977. After the withdrawal of the US from the UN Conference on the Law of the Sea in 1983 and President Reagan's declaration of a 200-mile Exclusive Economic Zone around US coasts, IOS successfully won a contract in 1984 to survey this newly acquired territory off the California coast. This was so successful that further contracts during the next seven years were obtained for surveys in the Gulf of Mexico, off the east coast of the USA, off Alaska, around the Aleutian Island Arc and around the Hawaiian Island chain. These surveys not only earned IOS very useful funds, national and international esteem, but also the Queen's Award for Technical Innovation.

The physical and engineering skills of NIO/IOS that were deployed in creating GLORIA were later used in the initial design and build of Autosub, an autonomous submersible capable of diving to several thousand meters in the ocean and traversing hundreds of kilometres. This vehicle was completed at SOC in 1998, and has since been deployed in many scientific missions observing and measuring in areas that were hitherto inaccessible. The design led the way for industry to design and build autonomous vehicles for commercial operations.

In the biological sciences, the research work on whales, initiated by Discovery Investigations, provided essential data to the International Whaling Commission which contributed to the eventual bans on commercial whaling of some stocks and species and the preservation of whale populations. Much of the work on the vertical and horizontal distributions of oceanic animals carried out at NIO and at IOS provided essential context and understanding needed to address problems presented by the possible use of the oceans for disposal of wastes of various types. This understanding was enhanced by the great expansion of benthic ecology which produced its own insights but also quantitative data on the biological link between the ocean-bottom sediments and the open ocean above them. The pelagic side of this ecological work has now largely been replaced by process-orientated studies of population sizes and biomass linked to models, and there is virtually no taxonomic work on midwater animals remaining at Southampton. But these older datasets still provide invaluable information on biodiversity and community structure and are used within international databases; they also have the potential to act as a baseline against which change – for example climate change – can be measured.

IOS built the new Joseph Proudman building at Bidston and also acquired new buildings in Hambledon, near to Wormley, in 1979. The biologists

The aftermath of the storm that coincided with the renaming of the Institute on 16 October 1987.

moved to Hambledon, but further plans for moving staff there were halted as national funding for science was squeezed in the early 1980s and financial stringency was imposed. In 1985 NERC, following its new Corporate Plan, decided to centralise the management of its various scientific laboratories and appoint Directors of Science to reside at the headquarters in Swindon. John Woods was appointed NERC Director of Marine Science and generated a new programme for the marine laboratories, described in 'The Challenge'.

Funds were very tight, particularly commissioned funds for applied wave research. For this reason, the Taunton Laboratory of IOS had to be closed in August 1985, the intention being to relocate the staff of the Sedimentation Group to Bidston and the Applied Wave Research staff to Wormley. However most staff members retired, moved to Royal Naval laboratories, universities and other organisations, or left to set up their own businesses. In 1987, NERC decided to separate once more the Bidston Laboratory of IOS from the Wormley Laboratory of IOS to become the Proudman Oceanographic Laboratory. IOS then became the IOS Deacon Laboratory (IOSDL), thereby reversing the decision in 1973 to merge the three laboratories!

The re-naming ceremony of IOSDL, conducted by Lord Sherfield, who had been a member of the House of Lords Committee, coincided with the Great Storm of 16 October 1987, which devastated the marquee and its scientific displays. But the celebration continued regardless with the new name being toasted amongst the wreckage of the displays with champagne drunk out of plastic cups!

Anthony Laughton retired in 1988 and the post of director was taken by Colin Summerhayes, who had formerly been at BP. Summerhayes was the last director of IOSDL during the reorganisation of the science programmes under John Woods and the preparations for the major upheaval and opportunities of the later move of the institute to Southampton.

Part of Woods' vision was global oceanography. The World Ocean Circulation Experiment (WOCE), planned during the 1980s, was to

The refitted Discovery *photographed in 2008. The bow section and foremast are the same as when built in 1962; so is the ship's call sign 'Golf' 'Lima' 'November' 'Echo'. Almost everything else has changed!*

be the first truly global-scale investigation of the role of the oceans in climate and a bid was made for the UK, through IOSDL, to play a major role. This required both new funding for research and an enhancement of the capabilities of RRS *Discovery*. Woods briefed the Prime Minister, Margaret Thatcher, on the issues and opportunities and as a result the funding duly materialised. In 1991 *Discovery* was completely rebuilt. Much of her hull was re-used but her engines and superstructure were replaced and she was lengthened by 3.25 m, thereby restoring her to her original design length. Her scientific complement was increased from 20 to 28 and crew reduced from 44 to 22. It is a tribute to her initial construction that she continues to operate after almost 50 years.

The new science funding was channelled into a newly-established James Rennell Centre (JRC) for Ocean Circulation, an offshoot of the IOSDL Marine Physics Group, established in 1990 at the University of Southampton's Science Park at Chilworth just north of Southampton. The international project office for WOCE remained at Wormley whilst the JRC was the first movement of staff to Southampton for what later became the Southampton Oceanography Centre (SOC).

In 1985, the House of Lords Select Committee on Science and Technology, with Henry Charnock as one of its advisors, had called for a general strengthening of contacts and collaboration between government-funded research institutes and higher education, and for strengthening in particular the links between Southampton University and IOSDL. This led to the creation of SOC as a joint venture between NERC and the University, bringing together in a purpose built facility the IOSDL from Wormley,

The end of an era. Tony Laughton (L) and David Webb (R) fold the Discovery *flag as the building at Wormley is vacated on 23 September 1995.*

the Research Vessel Services (then based at Barry), and the University's Departments of Oceanography and Geology. Funding for the construction of the new centre was approved by the Advisory Board for the Research Councils in 1989, and by the Universities' Funding Council in 1990.

A 13-acre site in Southampton Docks, alongside which research vessels could be moored, was leased from Associated British Ports and detailed planning for the new centre started at the beginning of 1989. Howard Roe co-ordinated the planning process with colleagues from NERC and the University and sixteen committees drew up plans for different parts and operations. These various committees all reported to a Management Committee chaired by John Woods and thence to the Project Board chaired alternately by the Chief Executive of NERC and the Vice Chancellor of the University. Planning took some three years, and inevitably ambition succumbed to financial realities with facilities such as a research plankton tower, towing tanks, an international conference centre, and a public display area all failing to make it off the drawing board. Despite such cuts, a magnificent building and campus was designed by the Culpin Partnership and built by Wimpey UK Ltd.

SOC was created to be a world-class centre and a national focus for all aspects of marine and earth sciences, for teaching and training in these areas, and for the development and transfer to industry of appropriate technology. It was also built to serve the national marine science community

Later demolition is underway to make room for luxury flats. (Photo courtesy of D.K. Poulter.)

by hosting facilities such as the fleet of NERC research vessels and the National Oceanographic Library which had been created at IOSDL in 1989 from the its own library. The total cost of the SOC project was some £50 million, a huge commitment to science and recognition of the importance of marine science to the UK.

The centre attracted attention from the outset, both because of its scale and status, but also because it was conceived as an experiment; could large-scale government-funded research facilities be successfully integrated with and managed by a University? Numerous presentations were made to a variety of audiences; a travelling road show – complete with model – toured the UK; the then-Prime Minister, John Major, visited a very muddy building site one rainy evening in 1992. The Minister for Science, William Waldegrave, visited in June 1993 and buried a time capsule built in the shape of Autosub – the hugely successful autonomous submarine developed at both Wormley and Southampton – beneath the front door, and the Mayor of Southampton topped the building out in June 1994.

SOC was completed in the summer of 1995, and staff and students moved in during the late summer and autumn of that year. Staff of the JRC were already in Southampton and moved down from Chilworth. The Duke of Edinburgh formally opened SOC in April 1996, and the old IOSDL closed. This was preceded by a party in the grounds at Wormley when current and retired staff, families and friends gathered for the

The Southampton (now National) Oceanography Centre with the refitted RRS Discovery *alongside.*

wake–during which the old flag was ceremoniously lowered by Tony Laughton. It was the end of an era at Wormley and the beginning of a new venture in a world that was very different from that when NIO was created in 1949. The buildings at Wormley were eventually demolished and replaced by expensive luxury residential accommodation, but the name has been immortalised! A new species of deep sea hag fish was named *Myxine ios* in honour of the institute. No sooner was this published then the name of the institute changed to SOC!

The first Director of SOC was Professor John Shepherd, who oversaw the establishment of the centre and began the process of bringing together the different communities from different employers, the wealth of equipment and infrastructure. The old discipline-based departments at Wormley had been reconfigured before the move, both to reflect the desire to promote the science from an interdisciplinary perspective but also to emphasise that this was a new beginning and not simply a transfer of existing structures into a new building. Within the centre the scientists from Wormley formed four research divisions–the James Rennell Division for Ocean Circulation and Climate (primarily the old physics department and the modellers); the George Deacon Division for Ocean Processes (biologists, physical and chemical oceanographers); the Challenger Division for Seafloor Processes (headed initially by Summerhayes, the last director of IOSDL, and comprising geology, geophysics and benthic biology); and the Ocean Technology Division (the instrument and workshop groups). These NERC divisions complemented the research of the University oceanography and geology departments, and from the outset staff in the divisions actively participated in student teaching and training.

Over time, the composition and focus of both the NERC divisions and the University Departments changed as integration developed between the various communities, and the centre expanded. Howard Roe became the second (and last!) Director of SOC in 1999. When he retired in 2005, there were some 1,119 staff and students at SOC, of whom 520 were undergraduates and 189 were postgraduates, and the total budget of the centre then was about £34 million. The scale and scope of SOC now far exceeded that of NIO and IOS and generated a large focal point for UK oceanography, for research, for education and for the provision of national infrastructure and facilities.

The creation of SOC was a huge success, but managerially it was always an unwieldy beast. There were continual tensions between NERC and the University and between their parent funding bodies, over how far the intended integration between the communities should proceed, and how the centre should be managed and owned. Following Roe's retirement, NERC and the University reconfigured the internal management, bringing back much of the infrastructure and activities into direct NERC control rather than continuing to manage the centre through the University. As with IOS in 1973 an earlier decision was reversed! Under the new (and current) Director, Ed Hill, the centre was renamed the National Oceanography Centre, Southampton, as a public declaration of its role and as such NOCS continues to be extremely successful.

NOCS is the proud legacy of Group W, NIO, IOS and IOSDL and of all the expert and devoted staff who had the privilege of working there. The national and international respect which NIO earned in the early days under Dr Deacon provided a firm foundation for its future growth and development. His insistence on maintaining the broad range of scientific disciplines within a single organisation and incorporating within this, very strong technical and engineering teams to work directly with the scientists proved to be essential for addressing the vital role of the oceans in issues as diverse as climate change, resources, pollution and socio-economics that continue to face us today. Many of the new technologies developed have subsequently been fed into industry.

There are very many scientists, engineers, technicians and support staff who worked at NIO and who remember the thrill of being at the cutting edge of science and the joy of working in a laboratory with such an informal atmosphere among friendly colleagues and under an inspiring director. Those of us who survive are grateful for having being part of it and for having the opportunity to describe it for future generations of oceanographers.

Annex 1

Cruises of RRS *Discovery II*

Major events	
1929	Built Ferguson Bros, Port Glasgow, Scotland
1929-31	First Commission
1933 March	Completes fifth ever circumnavigation of Antarctica
1950	*Discovery II* and *William Scoresby* purchased by Admiralty and given to NIO
1950	HM the King approved prefix "Royal Research Ship" Admiralty granted warrant to use Blue Ensign undefaced
1962	Scrapped

Cruises during the lifetime of NIO

Year	Dates	Station numbers	Science discipline	Chief Scientist	Working area/ports
1950	20.4–21.5	2653-2671	Marine biology	Mackintosh	Plymouth–Malta
	27.5–13.9	2672-2705	Marine biology	Herdman	Malta–Melbourne
	3.10–9.12	2706-2772	Marine biology	Herdman	Melbourne–Dunedin
	9.12 – 9.3.1951	2773-2816	Marine biology	Herdman	Dunedin–Sydney
1951	21.5–25.7	2817-2866	Marine biology	Herdman	Sydney–Simonstown
	7.8–6.12	2867-2911	Marine biology	Herdman	Simonstown – Plymouth
1952	29.5–15.6	2913-2936	Marine biology	Herdman	Plymouth–Plymouth
	26.6–6.8	2937-3039	Geology and geophysics	Herdman	Plymouth–W Approaches–NE Atlantic–Plymouth
	16.8–11.9	3040-3073	Marine biology	Mackintosh	Plymouth–Plymouth
	28.9–14.10	3074-3078	Electrode trials	?	Plymouth–Plymouth
	10.1952 – 3.1954		Laid up and refit		
1954	18.3–9.6	3092-3106	Marine biology	David	Plymouth–Azores – Plymouth
	17.6–5.7	3107-3140	Geology and geophysics	Hill (Cambridge)	Plymouth–Mid Atl. Ridge–Plymouth

Year	Dates	Station numbers	Science discipline	Chief Scientist	Working area/ports
	10.7–28.7	3141-3147	Marine biology	Currie	N Spain
	9.9–30.9	3162-3181	Marine biology	Mackintosh	W Portugal
	20.10 – 30.10	3182	Admiralty Research Lab		
	10.10 – 14.11	3183-3185	Geology and geophysics	Hill (Cambridge)	Plymouth–Gibraltar
	14.11 – 1.12	3186-3202	Marine biology	David	NE Atlantic – Plymouth
1955	25.1–26.1		Admiralty charter		Plymouth–Plymouth
	17.2–17.3		Air Ministry/Met Office		Plymouth–OWS Kilo–Plymouth
	March		Shipborne wave recorder trials		
	26.3–25.4	3211-3227	Marine biology and Admiralty Research Laboratory	Currie	Plymouth–Lerwick – Torshavn–Plymouth
	13.5–29.6	3230-3274	Marine biology + wave buoys and floats	David	Plymouth–Lisbon – Plymouth
	$3\frac{1}{2}$ months		Admiralty charter		
	7.9–3.10	3288-3332	Marine biology	Mackintosh	Plymouth–Canary Is–Plymouth
	11.10–22.11	3333-3363	Deep currents/ marine biology	Swallow	Plymouth–Gibraltar – Plymouth
1956	14.2–8.3		Current measurements	Swallow	Plymouth–Lisbon – Plymouth
	23.4–23.5		Current measurements	Swallow	Plymouth–Lerwick – Plymouth
	19.7–30.8	3403-3467	Geology and geophysics	Hill (Cambridge)	Plymouth–Azores – Mid Atl Ridge–Plymouth
	3 months		Admiralty charter		
	7.9–10.10	3468-3488	Marine biology	David	Plymouth–Azores – Plymouth
1957	1.2–29.4	3509-3548	Gulf Stream currents IGY sections 48°N	Herdman	Plymouth–Bermuda – Charleston–WHOI – Plymouth
	11.5–14.7		IGY sections	Herdman	Liverpool–Lerwick – Bergen–Trondheim
	$2\frac{1}{2}$ months		Admiralty charter		
	August		Geology and geophysics	Cambridge/ Liverpool	
	15.8–13.9	3587-3624 3625-3650	IGY sections 24°N 32°N	Herdman/ Swallow	Plymouth–Tenerife – Nassau–WHOI – Gibraltar–Plymouth
	26.9–18.12		Current measurement	Swallow	Plymouth–Galway – Plymouth
	12.1957–3.1958		Refit and ‘special’ survey		
1958	4.3–20.4	3652-3714	Marine biology	Currie	Plymouth–Canary Islands–Plymouth

Year	Dates	Station numbers	Science discipline	Chief Scientist	Working area/ports
	24.4–30.4	3715-3722	Trials–fish detection	Tucker	UK waters
	9.5–27.7	3723-3822	Currents, and geology and geophysics	Swallow/ Hill (Cambridge)	Plymouth–Lisbon – Lisbon–Plymouth
	14.8–16.9	3823-3846 3866-3884 3885-3901	IGY sections 58°N 46°N 43°N	Herdman	Europe to Mid-Atl Ridge, 60° and 43°N
	7.10–20.10		Trials–fish detection	Tucker	Plymouth–Plymouth
	1.11–12.12		IGY sections		Bay of Cadiz, Straits Gibraltar
1959	28.2–22.3	4012-4030	Trials–fish detection	Tucker	Off Scotland
	28.3–9.4		Survey		Plymouth–Plymouth
	5.5–25.6		Admiralty charter		Plymouth–Lerwick – Plymouth
	14.8–5.10	4228-4254	Marine biology	Currie	Madeira, Lisbon
	13.10–1.11		Admiralty charter		
	December		MAFF		
	9.12–13.12		Instrument trials	Laughton	Plymouth–Plymouth
1960	5.2–10.3		Trials NIO water bottle		Western Mediterranean
	18.3–11.4	4269-4290	Geology and geophysics	Laughton	Oporto, Galicia Bank
	20.4–13.5		Geology of shelf		Plymouth–Cork – Plymouth
	16.5–24.5		Aberdeen Fish Cruise	Stubbs	Plymouth–Aberdeen
	27.5–21.6		ICES Survey and "Overflow -60"		Aberdeen – Vesterhaven–Belfast – Plymouth Faroe –Iceland Ridge
	1.7–4.8		Admiralty charter		Plymouth–Gibraltar – Plymouth
			Geology of shelf	?	Off Portugal
	22.8–3.10	4515-4547	Geology and geophysics	Hill (Cambridge)	Plymouth–St Nazaire –Plymouth–MA Ridge–Plymouth
	13.10–26.10		Instrumental trials	?	Falmouth–Biscay
	18.11–8.12		IGY sections	?	Plymouth–Gibraltar – Cadiz–Plymouth
1961	3.2–24.2		Geology of shelf	?	
	?		Instrument trials	?	
	Mar–May		Admiralty charter		
	21.6–28.7		Current measurements	Swallow	Plymouth–Torshavn – Lerwick–Plymouth
	14.8–17.10	4668-4770	Marine biology	David	Plymouth–Azores – Canary Is–Madeira – Plymouth

Year	Dates	Station numbers	Science discipline	Chief Scientist	Working area/ports
	4.11–27.11		Instrument trials	Cartwright	
1962	4.1–10.3		Geology and geophysics	Hill (Cambridge)	Plymouth–Tenerife – Plymouth
	22.3–18.4	4825-4842	Marine biology	?	Plymouth–Madeira – Plymouth
	26.4–17.5		Shelf geology	Stride	SW British Isles
	24.5–4.6	4916-4963	Marine biology	?	Plymouth–Tenerife – Plymouth
1961	3.2–24.2		Geology of shelf	?	Plymouth–Cork – Plymouth
	?		Instrument trials	?	
	Mar–May		Admiralty charter		
	21.6–28.7		Current measurements	Swallow	Plymouth–Torshavn – Lerwick–Plymouth
	14.8–17.10	4668-4770	Marine biology	David	Plymouth–Azores – Canary Is–Madeira – Plymouth
	4.11–27.11		Instrument trials	Cartwright	
1962	4.1–10.3		Geology and geophysics	Hill (Cambridge)	Plymouth–Tenerife – Plymouth
1962	22.3–18.4	4825-4842	Marine biology	?	Plymouth–Madeira – Plymouth
1962	26.4–17.5		Shelf geology	Stride	SW British Isles
1962	24.5–4.6	4916-4963	Marine biology	?	Plymouth–Tenerife – Plymouth

Annex 2

Cruises of RRS *Discovery*

Major events	
1961/62	Built by Hall Russell, Aberdeen
21.3.1962	Launched
31.1.1963	Named by Lady Hailsham, Pool of London
10.1968–1.1969	Major refit by Hall Russell, Aberdeen. Internal layout altered. IBM 1800 computer installed
10.1990–6.1992	Refit in Viana do Castelo, Portugal. Included lengthening, new engines and internal layout

Cruises during the lifetime of NIO					
Year	**Cruise**	**Dates**	**Science discipline**	**Chief Scientist**	**Working Area**
1963	1	1.6–20.8	Physics, biology, chemistry	Currie	South East Arabian Upwelling
	2	23.8–4.12	Geology, geophysics, chemistry, physics	Hill (Cambridge)	NW Indian Ocean, passage to UK
1964	3	15.2–28.9	Physics, biology, chemistry	Swallow, Currie	W Indian Ocean, passages to and from UK
1965	4	2.2–1.4	Geology, geophysics	Hill (Cambridge)	NE Atlantic, Azores
	5	28.4–23.5	Trials, physics, geophysics	Tucker	NE Atlantic shelf edge
	6	1.6–1.7	Physics	Crease	NE Atlantic Faroe Bank
	7	7.7–6.8	Geology	Stride	NE Atlantic Faroes–Spain
	8	1.9–1.12	Biology	Currie	NE Atlantic Madeira, Canary Islands
1966	9	1.2–6.2	Trials	Herdman	NE Atlantic SW Approaches
	10	15.2–6.4	Physics	Swallow	NE Atlantic Madeira
	11	15.4–21.5	Geophysics, geology	Laughton	NE Atlantic Biscay, Azores
	12	31.5–21.6	Geology	Stride	NE Atlantic SW Approaches
	13	4.7–11.8	Physics		NE Atlantic Madeira

Year	Cruise	Dates	Science discipline	Chief Scientist	Working Area
	14	1.9–30.9	Geology, geophysics	(Liverpool, Bristol, Durham)	NE Atlantic
	15	7.10–28.11	Biology	Clarke	NE Atlantic Azores, Canary Islands
1967	16	20.1–6.5	Geology, geophysics, GLORIA	Laughton	Red Sea, Gulf of Aden, NW Indian Ocean
	17	3.6–8.7	Physics, trials	Crease	NE Atlantic Faroes, Biscay
	18	14.7–23.8	Physiology, biology	Clarke	NE Atlantic Canary Is
	19	12.9–12.10	Geology, geophysics	Stride	English Channel, NE Atlantic
	20	20.10–18.12	Physics	Swallow	NE Atlantic Biscay
1968	21	14.1–1.4	Biology, physics	David	NE Atlantic NW Africa
	22	18.7–25.7	Physics (tides)	Cartwright	NE Atlantic St Kilda
	23	31.7–19.8	Geophysics	Laughton	NE Atlantic Biscay
	24	19.9–30.9	Geology, geophysics	Tucker	NE Atlantic Rockall, Moray Firth
1969	25	25.1–28.3	Physics (MEDOC 1969)	Swallow	NW Mediterranean
	26	3.4–9.5	Physics, chemistry	Bowden (Liverpool)	NE Atlantic NW Africa
	27 & 28	9.6–24.7	Geology, geophysics, biology	Stride	NE Atlantic W Mediterranean
	29	2.8–6.10	Geophysics	Laughton	NE Atlantic Azores, Rockall
	30	15.10–15.12	Biology, physics	David	NE Atlantic NW Africa
1970	31	21.1–12.3	Physics (MEDOC 1970)	Swallow	NW Mediteranean, Gulf Cadiz
	32	24.3–7.4	Equipment trials	Crease	NE Atlantic Biscay
	33	14.4–21.5	Geophysics, geology	Laughton	NE Atlantic Rockall, King's Trough
	34	2.6–29.6	Physics, meteorology (JASIN)	Swallow	NE Atlantic Biscay
	35	11.7–19.8	Geology, geophysics, biology	Stride	W Mediterranean, Gulf Cadiz
	36	26.9–13.11	Biology, physics	David	NE Atlantic Azores
	37	25.11–10.12	Physics, meteorology	Watson	NE Atlantic Biscay
1971	38	22.1–7.4	Physics	Swallow	NE Atlantic Biscay, Gulf Cadiz
	39	21.4–7.6	Biology	Foxton	NE Atlantic Iceland Basin, Rockall
	40	19.6–26.7	Geology, physics	Stride	E Mediterranean, NE Atlantic
	41	9.8–19.9	Physics (tides), meteorology	Cartwright, Watson	NE Atlantic Hebrides, Shetland
	42	21.9–30.9	GLORIA, fishing	Rusby	Sea of Hebrides
	43	7.10–25.11	Geophyics, geochemistry	Laughton	NE Atlantic Azores, Gib. Ridge
	44	4.12–13.12	Physics	Swallow	NE Atlantic Biscay

Year	Cruise	Dates	Science discipline	Chief Scientist	Working Area
1972	45	4.2–16.4	Biology, physics	David	NE Atlantic NW Africa
	46	27.4–30.5	Physics, Meteorology	Tucker, Swallow	NE Atlantic Rockall, Biscay
	47	9.6–3.7	Geology, geophysics	Roberts	NE Atlantic Rockall
	48	12.7–22.8	Physics, biology, chemistry	Bowders (Liverpool)	NE Atlantic NW Africa
	49	31.8–2.10	JASIN 1972 Physics, meteorology	Swallow	NE Atlatnic Iceland Basin, Rockall
	50	11.10–26.10	Equipment trials, geology	Stride	N Atlantic, 32°N section, Bermuda
	51	4.11–18.12	Physics, geology	Crease, Swallow	N Atlantic, 32°N section, Bermuda
1973	52	17.2–28.3	Physics, biology	Swallow, David	N Atlantic, 32°N section, Bermuda
	53	1.4–14.6	Physics (MODE-1)	Swallow	Bermuda, N Atlantic

Annex 3

Acronyms and abbreviations

AEC	Atmospheric Environment Committee (of the UK Aeronautical Research Council)
APG	Applied Physics Group (NIO)
ARL	Admiralty Research Laboratory (Teddington, Middlesex)
ASDIC	Allied Submarine Detection Investigation Committee (early UK name for SONAR)
BANZARE	British Australian (and) New Zealand Antarctic Research Expedition (1929-31)
BEM	British Empire Medal
BODS	British Oceanographic Data Service
^{14}C	Radioactive Carbon tracer used for primary production studies
CAT	Clear air turbulence
CBE	Commander of the most excellent order of the British Empire
CIA	Central Intelligence Agency (USA)
CITES	Convention on International Trade in Endangered Species
CMOS	Complementary metal–oxide–semiconductor
CMT	Committee on Marine Technology (UK)
CTD	Conductivity-temperature-depth probe
DAMTP	Department of Applied Mathematics and Theoretical Physics (Cambridge)
DB1 (2,3)	Large data buoys (NIO)
DNC	Department of Naval Construction (UK)
DSDP	Deep Sea Drilling Project
DSL	Deep scattering layer
DSO	Distinguished Service Order, military decoration.

DRCM	Direct-reading current meter
DV	Drilling Vessel
FAMOUS	Franco-American Mid-Ocean Undersea Study
FBA	Freshwater Biological Association, Windermere.
FRS	Fellow of the Royal Society
FRSE	Fellow of the Royal Society of Edinburgh.
GARP	Global Atmospheric Research Program
G&G	Department of Geodesy and Geophysics (Cambridge)
GEBCO	General Bathymetric Chart of the Oceans
GEK	Geomagnetic-electrokinetograph
GLORIA	Geological Long Range Inclined ASDIC
HO	Hydrographic Office (of the Admiralty)
IAPSO	International Association for the Physical Sciences of the Oceans
ICES	International Council for the Exploration of the Sea
ICOT	Institute of Coastal Oceanography and Tides (UK)
IGY	International Geophysical Year (1957-8)
IIOE	International Indian Ocean Expedition (1959-65)
IKMT	Isaacs Kidd midwater trawl
IOC	Intergovernmental Oceanographic Commission (of UNESCO)
IOS	Institute of Oceanographic Sciences (1973 – 1987)
IOSDL	Institute of Oceanographic Sciences Deacon Laboratory (1987-1994)
IPOD	International program on Ocean Drilling
IWC	International Whaling Commission
JASIN	Joint Air-Sea Interaction Experiment
JOIDES	Joint Oceanographic Institutions for Deep Earth Sampling (group)
JONSWAP	Joint North Sea Wave Project
JPOTS	Joint Panel on Oceanographic Tables and Standards
JRC	James Rennell Centre (component of IOSDL)
KNMI	Koninklijk Nederlands Meteorologisch Instituut
LORAN	LOng RAnge Navigation

MAFF	Ministry of Agriculture Fisheries and Food
MATSU	Marine Technology Support Unit (of the CMT)
MBA	Marine Biological Association (Plymouth, UK)
MBL	Marine Biological Laboratory (in Woods Hole)
MIAS	Marine Information Advisory Service (UK)
MDD	Mine Design Department (of the Admiralty)
MODE	Mid-Ocean Dynamics Experiment (joint US-UK)
MSES	Marine Scientific Equipment Service (UK)
MV	Motor Vessel
N50V	Vertically hauled plankton net with a 50 cm mouth diameter
N70V	Vertically hauled plankton net with 70 cm mouth diameter.
NASA	National Aeronautics and Space Administration (USA)
NERC	Natural Environment Research Council
NHM	Natural History Museum (London)
NIO	National Institute of Oceanography
NOCS	National Oceanography Centre, Southampton
NOL	National Oceanographic Library
NPL	National Physical Laboratory (UK)
NSTIC	Naval Science and Technology Information Centre (UK)
NUWC	Naval Underwater Weapons Center (USA)
OBS	Ocean Bottom Seismograph
OECD	Organisation for Economic Co-operation and Development.
ONR	Office of Naval Research (USA)
PES	Precision echosounder
PUBS	Pop-Up Bottom Seismograph
RMS	Royal Mail Ship
RMT1+8	Rectangular Midwater Trawl combining nets of 1 and 8m^2 mouth areas
RNSS	Royal Naval Scientific Service
RRS	Royal Research Ship
SAC	Scientific Advisory Committee (to the British War Cabinet)
SBWR	Shipborne Wave Recorder
SCAR	Scientific Committee on Antarctic Research
SCODS	Standing Committee on Ocean Data Stations (UK)

SCOR	Scientific Committee on Oceanic Research (initially Special Committee)
SIO/Scripps	Scripps Institution of Oceanography (La Jolla, USA)
SMBA	Scottish Marine Biological Association, Oban.
SMRU	Sea Mammal Research Unit (Successor of WRU)
SOC	Southampton Oceanography Centre (now NOCS)
SOFAR	Sound fixing and ranging
SONAR	Sound navigation and ranging
SOND	September, October, November, December (1965) Cruise 8 of RRS *Discovery*
UCS	Unit of Coastal Sedimentation (UK)
ULTI	University of Liverpool Tidal Institute
UNCLOS	United Nations Conference (later Convention) on the Law of the Sea
UNESCO	United Nations Educational, Scientific and Cultural Organization
WHOI	Woods Hole Oceanographic Institution
WREN	Women's Royal Naval service.
WRU	Whale Research Unit (of NIO, based at the Natural History Museum London)

Annex 4

Author biographies

Martin Angel joined NIO after participating in the IIOE as a John Murray Research Student. He continued the research on zooplankton ecology he had initiated during the IIOE, taking a special interest in an important but poorly known group, the planktonic ostracods. He published over 150 scientific papers and edited a Festschrift volume of *Deep-Sea Research* entitled *A Voyage of Discovery*, to celebrate Sir George Deacon's 70th birthday. He was then appointed a co-editor of *Progress in Oceanography*, a position he held for the next 25 years. He succeeded Peter David as head of the Biology Group in 1985. After retiring in 1997, he has continued to serve on a variety of committees, was lead writer of the 2000 OSPAR review of the state of the open Atlantic, and still continues to write papers and develop web-based plankton atlases.

David Cartwright was born in London in 1926. After graduating in the Natural Sciences Tripos at Cambridge he took a second degree in honours mathematics at King's College London. He joined the NIO in 1954, where he took special interest in wave motions and their statistics. This interest later embraced storm surges, tidal motions, and satellite altimetry. At NIO he rose to the rank of Special Merit SPSO, and was appointed Assistant Director of IOS with responsibility for the laboratory at Bidston, near Liverpool. Starting in 1963-65 and on later occasions he worked with Walter Munk at SIO in La Jolla. In 1984 he was elected a Fellow of the Royal Society. He has authored more than 80 scientific papers, and a book *Tides: A Scientific History*.

Jim Crease was one of the first scientists to join the newly-founded NIO, having come first as a vacation student. Initially researching storm surge theory and ocean currents, he first worked closely with Henry Charnock and then John Swallow and was a major contributor to NIO research in the area of marine physics. He was leader of the Marine Physics Group from 1974-86. The introduction of satellite navigation in 1968 led on to a broader interest in data management, particularly in an international oceanographic context. His retirement from IOS in 1988 was followed by 18 years' collaboration with Ferris Webster at the University of Delaware on the management of data from the World Ocean Circulation Experiment. He was born in 1928.

Fred Culkin graduated in chemistry (London University, External) in 1952 and received his London External PhD in 1960 for research on trace elements in seawater carried out at Liverpool University. He joined NIO in 1960 to work with Roland Cox in a study of the chemical and physical properties of seawater and later took over responsibility for this project following the untimely death of Cox in 1967. This work contributed to the development of the Practical Salinity Scale 1968 and the equation of state of seawater 1980. In 1968 he became Director of the IAPSO Standard Seawater Service operating from NIO. His other interests were salinity standards and the organic chemistry of marine organisms and sediments.

Margaret Deacon read history at university without any intention of becoming a historian of oceanography. This came about when her father, George Deacon, asked her to look at some references he lacked time to follow up. These references were to observations by early Fellows of the Royal Society during the 1660s–a time immediately following the period she had recently been studying for her special subject, and even involving some of the same personalities, but in very different roles. This eventually led in 1971 to the publication of *Scientists and the Sea, 1650-1900*, which is still in print and regarded as a classic pioneering work in the field, later recognised by the award of an honorary doctorate from Dalhousie University. In 1980 she was awarded a Hartley Fellowship in the Department of Oceanography of the University of Southampton. Following her marriage to David Seward in 1982 she continued to be attached to the department until they moved to the West Country in 2001. Her earlier publications dealt mainly with topics pre-1900 but following the death of her father in 1984 she undertook the cataloguing of official files from his time as Director of NIO and associated papers (now at NOL). During this exercise many of the documents that form the basis of the account of the foundation of NIO given in this book were uncovered.

Peter Foxton, a biologist, joined NIO in October 1949 and sailed on the 1950-51 commission of RRS *Discovery II* to the Southern Ocean. His initial research stemmed from this experience and concerned the relationship between the life history and distribution of salp species, while his later studies focused on the vertical structure of the pelagic ecosystem with particular reference to deep-sea shrimps. In 1974 he joined the NERC Science Division, with responsibility for marine life sciences and polar sciences. He retired in 1987 and was appointed OBE in 1988.

John Gould joined NIO in 1967 following a first degree in maths and physics from King's College London and a PhD from the University College of North Wales, Bangor. He worked with John Swallow on measuring ocean currents and in 1986 he succeeded Jim Crease as head of Marine Physics. From 1994 he was International Director of, first, the World Ocean Circulation Experiment (WOCE) and later of the Climate Variability and Predictability Study of the World Climate Research Programme until his retirement from the Southampton Oceanography Centre in 2002. He then became Director of the international Argo project that operates 3,000 neutrally buoyant floats derived from the Swallow floats. He was born in 1942.

Anthony S. Laughton, born in 1927, joined NIO as a geophysicist in 1955, having obtained his doctorate in marine geophysics at Cambridge and then studied at the Lamont Geological Observatory in New York. At NIO he developed deep-sea underwater photography, prepared contoured charts of the ocean floor, studied the geological evolution of the Gulf of Aden and of the North Atlantic and exploited the use of long-range side-scan sonar (GLORIA), especially in revealing the tectonic structure of mid-ocean ridges. He led the UK into the international Deep Sea Drilling Project, and advised the government at the UN Conference on the Law of the Sea. From 1978-88 he was Director of IOS. In 1980 he was elected a Fellow of the Royal Society, and in 1987 awarded a knighthood for services to oceanography. He was awarded the Prince Albert 1st of Monaco Gold Medal for Oceanography (1980), the Founders Medal, RGS (1987) and the Murchison Medal, Geological Society of London (1989). After retirement he served as Chairman of GEBCO (General Bathymetric Chart of the Oceans) until 2003.

Michael Longuet-Higgins was born in 1925 at Lenham, near Maidstone, Kent. In 1945 he graduated at Cambridge University with a first-class degree in mathematics. From 1945 to 1948 he was a member of Group W, and in 1948 he returned to Cambridge to continue his research there. In 1951 he was awarded a Prize Fellowship at Trinity College. He spent 1951 to 1952 on a Commonwealth Fund Fellowship travelling and studying oceanography in the USA. He rejoined NIO in 1954. Elected F.R.S. in 1963, he is also a Foreign Associate of the US National Academy of Sciences. He has won many awards, including the Sverdrup Gold Medal of the American Meteorological Society. From 1967 to 1969 he assisted in establishing the School of Oceanography at Oregon State University. In 1969 he returned to England to take up a Royal Society Research Professorship held jointly at NIO and at the Department of Applied Mathematics at Cambridge, together with a Senior Research Fellowship at Trinity College. Since his nominal retirement in 1989 he has been a Research Physicist at the University of California, San Diego, becoming Emeritus in 1995. He is still active in science and mathematics.

Brian S. McCartney. Following a doctorate on electronic scanning of sonar receivers, and a couple of years in industry, he joined NIO in 1964 to study sound scattering from fish. In 1970 he was awarded the A.B. Wood medal of the Institute of Physics for underwater acoustics. He supported all branches of marine science with acoustic instrumentation and led the applied physics group from 1973 until he left the Institute in 1987 to become director of Proudman Oceanographic Laboratory at Bidston, from which he retired in 1998. He was born in 1938.

Howard Roe was born in 1943 and joined the Whale Research Unit of NIO in 1965, following a degree in zoology at University College London. He carried out research into methods of age determination in baleen whales and the feeding of Sperm whales, and spent one summer working at the whaling station in Iceland. He moved to NIO at Wormley in 1968 and thereafter investigated aspects of deep-sea biology, community structure and behaviour, and developed a range of sampling equipment and sensors. He was awarded a DSc in biological oceanography by UCL in 1998. He coordinated the development of the Southampton Oceanography Centre and became its Director from 1999 until retiring in 2005. As Director he co-founded the international Partnership for Observation of the Global Ocean (POGO) and chaired the development of the latest NERC research ship RRS *James Cook*.

Pauline Simpson joined the NIO in 1969 as the first appointed Chartered Librarian (with a degree in geology with oceanography). From 1985 she became Head of Information Services and took the NIO Library on a path to full automation, developing advanced services that resulted in the library being designated the 'National Oceanographic Library'. 1991-99 she was Chair of the UNESCO Intergovernmental Oceanographic Commission (IOC) Group of Experts on Marine Information Management and was instrumental in capacity building in developing countries and forging close data and information relationships worldwide. In 1998-99 she was President of the International Association of Aquatic and Marine Science Libraries and Information Centres (IAMSLIC). At Southampton from 1995, she was also a member of the University Library Senior Policy & Strategy Group and Faculty Leader for Engineering, Science and Mathematics. From 2001 she was a leader in the implementation of Digital Repositories, providing open access to research publications for Southampton University and the NERC and also IOC and IAMSLIC. She retired in November 2006.

Arthur Stride graduated in geology at Bristol University in 1950 and gained a DSc in 1969. He joined the Institute in1950. He demonstrated the importance of side-scan sonar for making geological maps, revealing the the effects of tidal and non-tidal currents on the continental shelf as well as showing the evolution of the continental slope and the floor of the eastern Mediterranean. He was author or co-author of over 80 papers and two books. He was awarded the Wollaston Fund of the Geological Society of London in 1967, and the Busk (gold) Medal of the Royal Geographical Society in 1976. He retired in 1982, but continued to publish papers until 1999.

Steve Thorpe was born in 1937. After graduating with an external degree in maths and physics from London University in 1958 during his National Service, he studied maths at Cambridge and joined the NIO in 1962, working mainly on internal waves, mixing and the application of acoustics to the study of the upper parts of the ocean. Following Henry Charnock's retirement from the Chair in 1986, he was appointed Professor of Oceanography at Southampton University, and is now an Emeritus Professor. He was President of the Royal Meteorological Society, 1990-92, and was elected a Fellow of the Royal Society in 1991. He was awarded the Oceanography Society/US Navy Walter Munk Award in 1997 for work on acoustical oceanography, and the Fridtjof Nansen Medal of the European Geophysical Society in 2000 for investigations of mixing processes in the ocean. Since 'retiring' in 1997, he has written two books on turbulence in the ocean published by Cambridge University Press, and co-edited the

Encyclopedia of Ocean Sciences with John Steele and Karl Turekian. He is currently an Honorary Professor in the School of Ocean Sciences at the University of Bangor.

Malcolm J. ('Tom') Tucker was born in December 1924. He took an honours degree in physics in 1944, and joined the newly-formed Group W at the Admiralty Research Laboratory late that year. While in Group W and NIO he concentrated on developing instruments. However he also took a wider interest in wave research, and when he became Assistant Director in charge of the IOS Taunton laboratory in 1973, this became his main professional interest. He retired in 1984. He has written a textbook entitled *Waves in Ocean Engineering*.

References

Chapter 1

1. Deacon, M.B. *Scientists and the sea, 1650-1900: a study of marine science* (Ashgate, Farnborough, 1971, repro. 1997).
2. Rozwadowski, H.M. *The sea knows no boundaries: a century of marine research under ICES* (Univ. Washington Press, 2003).
3. Mills, E.L. '"Physische Meereskunde": from geography to physical oceanography in the Institut für Meereskunde, Berlin, 1900-1935'. *Historisch-Meereskundliches Jahrbuch* **4**, 45-70 (1997).
4. Schlee, S. *The edge of an unfamiliar world: a history of oceanography* (Dutton, New York, 1973).
5. Deacon, M.B. 'Crisis and compromise: the foundation of marine stations in Britain during the late 19th century'. *Earth Sci. Hist.* **12**(1), 19-47 (1993).
6. Mills, E.L. *Biological oceanography: an early history, 1870-1960* (Cornell Univ. Press, 1989).
7. Deacon, M.B. 'G. Herbert Fowler (1861-1940): the forgotten oceanographer'. *Notes & Rec. R. Soc. Lond.* **38**(2), 261-296 (1984).
8. Hackmann, W. *Seek and strike: sonar, anti-submarine warfare and the Royal Navy, 1914-54* (HMSO, London, 1984).
9. Marsden, R. 'Expedition to investigation: the work of the Discovery Committee' in Deacon, M., Rice, A.L. & Summerhayes, C. *Understanding the oceans: a century of ocean exploration* (UCL Press, London, 2001).
10. Coleman-Cooke, J. Discovery II *in the Antarctic: the story of British research in the Southern Seas* (Odhams, London, 1963).

Chapter 2

1. Deacon, G.E.R. 'Hans Pettersson, 1888-1966'. *Biog. Mem. Fellows R. Soc.* **12**, 405-21 (1966).
2. TNA CAB 90/4. Minutes of meetings of the Scientific Advisory Committee to the War Cabinet (1943).
3. Deacon, G.E.R. 'John Augustine Edgell, 1880-1962'. *Biog. Mem. Fellows R. Soc.* **9**, 88-90 (1963).
4. HO H02455/43. Proudman to Edgell (16 August 1943).
5. Charnock, H. 'George Edward Raven Deacon, 1906-1984'. *Biog. Mem. Fellows R. Soc.* **31**, 111-142 (1985).
6. Howarth, R.J. 'Sir Edward Crisp Bullard (1907-1980)' in *Oxford Dictionary of National Biography* (Oxford Univ. Press, 2004).

7. Lee, A.J. *The Directorate of Fisheries Research: its origins and development.* MAFF, Lowestoft, (1992). A copy of Tait's memorandum is in NOL GERD M3/9 (1992).
8. HO H02455/43. Taylor to Havelock (14 January 1944).
9. TNA CAB 90/4 (206-8). Memorandum by Admiral Edgell (1943).
10. NOL GERD M3/10. Proudman, J., Proposed National Institute for Oceanography (1 December 1943).
11. NOL GERD M3/11. Carruthers, J.N., Memorandum on post-war oceanography in Britain (17 January 1944).
12. NOL GERD M 3/13. Deacon, G.E.R., Memorandum on "Oceanographical Research" (1944).
13. HO H02455/43. Edgell to Carruthers (21 January 1944).
14. NOL GERD M3/12. Minutes of the meeting of the oceanography subcommittee of the National Committee for Geodesy and Geophysics (1 March 1944).
15. NOL GERD M3/3. Minutes of the meeting of the oceanography subcommittee of the National Committee for Geodesy and Geophysics (26 May 1944).
16. Charnock, H. 'Sea and swell forecasting for operational planning' in *Meteorology and World War II* (ed. Giles, B.D.) (Royal Meteorological Society, 1987).
17. Schlee, S. *The edge of an unfamiliar world: a history of oceanography* (Dutton, New York, 1973).
18. TNA CO 78/219/7. Discovery Committee and Scientific Subcommittee minutes (June 1944-December 1945) and CO 78/221/7 Discovery Committee, future prospects (1944-October 1945).
19. TNA CAB 90/6. Memorandum by the Treasury, discussed at the SAC meeting of 26 July 1946.
20. TNA ADM 116/5715. The National Institute of Oceanography. This contains details of the negotiations between the Admiralty and the Treasury, including Tizard's letter to the Admiralty of 27 August 1948.
21. Lowe. R. 'Sir (James) Alan Noel Barlow, 1881-1968' in *Oxford Dictionary of National Biography* (Oxford Univ. Press, 2004).
22. *Annual Report of the National Institute of Oceanography, 1949-50.*
23. Rice, A.L. 'Forty years of land-locked oceanography; the Institute of Oceanographic Sciences at Wormley'. *Endeavour* **18**, 137-146 (1994).
24. Deacon, M.B. 'List of the Personal and Scientific Papers of Sir George Deacon FRS'. *IOS Rep.* **301** (1992).

Chapter 3

1. Deacon, G.E.R. Notes for the Royal Society (personal manuscript).
2. Deacon, G.E.R. 'A general account of the Hydrology of the South Atlantic Ocean'. *Discovery Rep.* **7**, 171-238 (1933).
3. Coleman-Cooke, J. Discovery II *in the Antarctic: the story of British research in the Southern Seas* (Odhams, London, 1963).
4. Deacon, G.E.R. 'The Hydrology of the Southern Ocean'. *Discovery Rep.* **15**, 1-124 (1937).
5. Mills, E.L. 'From Discovery to discovery: the hydrology of the Southern Ocean, 1885-1937'. *Arch. Nat. Hist.* **332**(2), 246-264 (2005).
6. Deacon, G.E.R.. *The Antarctic Circumpolar Ocean*, 180pp (Cambridge Univ Press, NY, 1985).
7. Charnock, H. 'George Edward Raven Deacon 1906-1984'. *Biog. Mem. Fellows R. Soc.* **31**, 111-142 (1985).

Chapter 4

1. Ursell, F. *Reminiscences of the early days of the spectrum of ocean waves* in *Wind-over-Waves Couplings* (eds Sajjadi, S.G., Thomas, N.H. & Hunt, J.C.R.) (Oxford Univ. Press, 1999).
2. Lamb, H. *Introduction to Hydrodynamics*, 6th ed. (Cambridge Univ. Press, 1932).
3. Ursell, F. Analysis of sea waves. Admiralty Research Laboratory Rep. A.R.L./103.30/R.I./W. (19 March 1945).
4. Barber, N.F., Ursell, F., Darbyshire, J. & Tucker, M.J. 'A frequency analyser used in the study of ocean waves'. *Nature, Lond.* 329-335 (1946).
5. Barber, N.F., Ursell, F. 'The generation and propagation of ocean waves and swell. I. Wave periods and velocities'. *Phil. Trans. R. Soc. Lond.* A 240, 527-560 (1948).
6. Darbyshire, J. 'The generation of waves by wind'. *Proc. R. Soc. Lond.* A 215, 299-328 (1952).
7. Deacon, M.B. (ed.) *Oceanography, Concepts and History* (Stroudsberg, PA., Dowden, Hutchinson & Ross, 1978).
8. Darbyshire, J. 'Reminiscences on the early days of wave research'. *Ocean Challenge* 12, 23-30 (2003). The section written by Jack Darbyshire in Chapter 4 is taken from this article.
9. Darbyshire, J. & Tucker, M.J. 'A frequency analyser used in the study of ocean waves'. *Nature* 158, 329-335 (1946).
10. Barber, N.F. 'The behaviour of waves in tidal streams'. *Proc. R. Soc. Lond.* A 198, 81-93 (1949).
11. Deacon, G.E.R. 'Storm warnings from waves and microseisms'. *Weather* 4, 74-79 (1949).
12. Darbyshire, J. 'Identification of microseism activity with sea waves'. *Proc. R. Soc. Lond.* A 202, 439-448 (1950).
13. Mortimer, C.H. *Looking back sixty-four years to my first encounter with waves* (Unpublished memoir, 2006).
14. Barber, N.F. *Water waves*. Wykenham Publications (Taylor and Francis). London (1989).
15. Carson, Rachel. *The Sea around Us*. (Oxford Univ. Press 1961).
16. Tucker, M.J. 'Sea wave recording'. *Dock Harb. Author.* 34, 207-210 (1953).
17. Cherry, D.W. & Stovold, A.T. 'Water movements and earth currents: electrical and magnetic effects'. *Nature* 161, 192-193 (1948).
18. Bowden, K.F. 'Measurement of wind currents in the sea by the method of towed electrodes'. *Nature* 171, 735-737 (1953).
19. Longuet-Higgins, M.S. & Ursell, F. 'Sea waves and microseisms'. *Nature* 162, 700 (1948).
20. Kedar, S., Longuet-Higgins, M.S., Webb, F., Graham, N., Clayton, R. & Jones, C. 'The origin of deep ocean microseisms in the North Atlantic'. *Proc. R. Soc. A–Math. Phys. Eng. Sci.* 464, 777-793 (2008).
21. Cooper, R.I.B. & Longuet-Higgins, M.S. 'An experimental study of the pressure oscillations in standing water waves'. *Proc. R. Soc. Lond.* A 206, 424-435 (1951).
22. Longuet-Higgins, M.S. 'Can sea waves cause microseisms?' *Proc. Symp. Microseisms. US Nat. Acad. Sci. Publ.* 306, 74-93 (1952).
23. Longuet-Higgins, M.S. 'On the statistical distribution of the heights of sea waves'. *J. Mar. Res.* 11, 245-266 (1952).

Chapter 5

1. Hardy, A. Appendix in *Great Waters* (Collins, London, 1967). (This book lists all Discovery Reports written between 1929 and April 1967 in which the work described here, and much more besides, was published.)
2. David, P.M. 'The photography of live oceanic plankton animals'. *Int. Photo Technik* 1, 40-42, (1963).
3. Mackintosh, N.A. 'The voyage of *Discovery II*'. *Nature* 169, 52-53 (1952).
4. Baker, A.de C. 'Underwater photographs in the study of oceanic squid'. *Deep-Sea Res.* 4, 126-129 (1957).
5. Clarke, M.R. 'The identification of cephalopod beaks and the relationship between beak size and total body weight'. *Bull. Brit. Mus. Nat. Hist. (Zool.)* 8, 422-480, (1962).
6. Currie, R.I. & Foxton, P. 'The Nansen closing method with vertical plankton nets'. *J. Mar. Biol. Ass. UK* 35, 483-492 (1956).
7. Currie, R.I. & Foxton, P. 'A new quantitative plankton net'. *J. Mar. Biol. Ass. UK* 36, 17-32. (1957).
8. David, P.M. 'The neuston net: a device for sampling the surface fauna of the ocean'. *J. Mar. Biol. Ass. UK* 45, 313-320 (1965).
9. Currie, R.I., Fisher, A.E. & Hargreaves, P.M. 'Arabian Sea Upwelling' in *The Biology of the Indian Ocean* (eds Zeitzschel, B. & Gerlach, S. A. (Springer, Berlin, 1973).
10. Foxton, P. 'SOND Cruise 1965: Biological sampling methods and procedures'. *J. Mar. Biol. Ass. UK* 49, 613-620 (1969).
11. Badcock, J. 'The vertical distribution of mesopelagic fishes collected on the SOND cruise'. *J. Mar. Biol. Ass. UK* 50, 1001-1044 (1970).
12. Angel, M.V. 'Planktonic ostracods from the Canary Islands region: their depth distributions, diurnal migrations and community organization'. *J. Mar. Biol. Ass. UK* 49, 515-553 (1969).
13. Roe, H.S.J. 'The vertical distributions and diurnal migrations of calanoid copepods collected on the SOND cruise 1965: the total population and general discussion'. *J. Mar. Biol. Ass. UK* 52, 277-314 (1972).
14. Clarke, M.R. 'A new midwater trawl for sampling discrete depth horizons'. *J. Mar. Biol. Ass. UK* 49, 945-960 (1968).
15. Baker, A.de C., Clarke, M.R. & Harris M.J. 'The NIO combination net (RMT1+8) and further developments of rectangular midwater trawls'. *J. Mar. Biol. Ass. UK* 53, 167-184 (1973).
16. Herring, P.J. 'Bioluminescence in marine organisms'. *Nature* 267, 788-793 (1977).
17. Angel, M.V. & Fasham M.J.R. 'SOND cruise 1965: factor and cluster analysis of the plankton results, a general summary'. *J. Mar. Biol. Ass. UK* 53, 185-231 (1973).
18. Roe, H.S.J. 'Observations on the diurnal vertical migrations of an oceanic animal community'. *Mar. Biol.* 28, 99-113 (1974).

Chapter 6

1. Hardy, A. Appendix in *Great Waters* (Collins, London,1967). (This book lists all Discovery Reports written between 1929 and April 1967 in which the work described here, and much more besides, was published.)
2. Mackintosh, N.A. *The Stocks of Whales* (Fishing News (Books) Ltd., London, 1965).
3. Clarke, R. 'A great haul of ambergris'. *Nature* **174**, 155-156 (1954).

4. Clarke, R. 'A giant squid swallowed by a Sperm Whale'. *Norsk Hvalfangst-Tidende* **44**, 589-593 (1955).
5. Brown, S.G. 'International co-operation in Antarctic whale marking, 1960 to 1965'. *Norsk Hvalfangst-Tidende* **55**, 89-96 (1966).
6. Brown, S.G. 'Swordfish and Whales'. *Norsk Hvalfangst-Tidende*, **49**, 345-351 (1960).
7. Clarke, M.R. 'A review of the systematics and ecology of oceanic squids' in *Advances in Marine Biology, Vol. 4* (ed. Russell, F.S.) (Academic Press, London, 1966).
8. Clarke, M.R. 'The function of the spermaceti organ of the sperm whale'. *Nature* **228**, 873-874 (1970).
9. Clarke, R., Aguayo, L.A. & Paliza, O. 'Sperm whales of the Southeast Pacific. Part 1.1, Size range, external characters and teeth'. *Hvalradets Skrifter* **51**, 80pp. (1968).
10. Clarke, R. & Paliza, O. 'Sperm whales of the Southeast Pacific. Part 1. II, Morphometry'. *Hvalradets Skrifter* **53**, 106pp. (1972).
11. Crisp, D.T. 'The tonnages of whales taken by Antarctic Pelagic operations during twenty seasons and an examination of the Blue Whale Unit'. *Norsk Hvalfangst-Tidende* **10**, 389-393 (1962).
12. Bannister, J.L. & Gambell, R. 'The succession and abundance of fin, sei and other whales off Durban'. *Norsk Hvalfangst-Tidende* **54**, 45-60 (1965).
13. Gambell, R. & Grzegorzewska, C. 'The rate of lamina formation in sperm whale teeth'. *Norsk Hvalfangst-Tidende* **57**, 117-121 (1967).
14. Gambell, R. 'Seasonal cycles and reproduction in sei whales of the southern hemisphere'. *Discovery Reports* **35**, 31-134 (1968).
15. Gambell, R. 'Sperm whales off Durban'. *Discovery Reports* **35**,199-358 (1972).
16. Roe, H.S.J. 'Seasonal formation of laminae in the ear plug of the fin whale'. *Discovery Reports* **35**, 1-30 (1967).
17. Lockyer, C. 'The age at sexual maturity of the southern fin whale (*Balaenoptera physalis*) using annual layer counts in the ear plug'. *J. du Conseil* **34**, 276-294 (1972).
18. Roe, H.S.J. 'The food and feeding habits of the sperm whale (*Physeter catodon* L.) taken off the west coast of Iceland'. *J. du Conseil* **33**, 93-102 (1969).

Chapter 7

1. Sverdrup, H.U., Johnson, M.W. & Fleming, R.H. *The Oceans: Their Physics, Chemistry, and General Biology* (Prentice-Hall, New York, 1942).
2. Burling, R.W. *Generation of waves on water* (PhD Thesis, Univ. London, 1955).
3. Charnock, H. 'Wind stress on a water surface'. *Q.J.R. Met. Soc.* **81**, 639-640 (1955).
4. Bowden, K.F. 'The direct measurement of subsurface currents in the ocean'. *Deep-Sea Res.* **2**, 33-47 (1954).
5. Ritchie, G.S. *"Challenger": The life of a Survey Ship* (Hollis & Carter, London, 1957).
6. Swallow, J.C. 'A neutral-buoyancy float for measuring deep currents'. *Deep-Sea Res.* **3**, 74-81 (1955).
7. Stommel, H. 'Direct measurements of subsurface currents'. *Deep-Sea Res.* **2**, 284-285 (1955).
8. Stommel, H. 'The westward intensification of wind-driven ocean currents'. *Trans. Am. Geophysical Union* **29**, 202-206 (1949).
9. Warren, B.A. & Wunsch, C. *Evolution of Physical Oceanography* (MIT Press, Cambridge, 1981).

10. Stommel, H. 'The abyssal circulation'. *Deep-Sea Res.* **5**, 80-82. (1958).
11. Swallow, J.C. & Worthington, L.V. 'An observation of a deep counter-current in the western North Atlantic'. *Deep-Sea Res.* **8**, 1-19 (1961).
12. Longworth, H.R. & Bryden, H.L. 'Discovery and quantification of the Atlantic Meridional Overturning Circulation: the importance of 25°N' in *Ocean circulation: mechanisms and impacts - past and future changes of meridional overturning* (eds Schmittner, A., Chiang, J.C.H. & Hemming, S.R.) (AGU Geophysical Monograph 173, 2007).
13. Crease, J. 'Velocity measurements in the Deep water of the Western North Atlantic'. *J. Geophysical Res.* **67**, 3173-3176 (1962).
14. Swallow, J.C. 'The Aries current measurements in the Western North Atlantic'. *Phil. Trans. R. Soc. A* **270**, 451-460 (1971).
15. Crease, J. 'The flow of Norwegian Sea water through the Faeroe Bank Channel'. *Deep-Sea Res.* **12**, 143-150 (1965).
16. Swallow, J.C. & Bruce, J.G. 'Current measurements off the Somali coast during the southwest monsoon of 1964'. *Deep-Sea Res.* **13**, 861-888 (1966).
17. Swallow, J.C. 'The equatorial undercurrent in the western Indian Ocean'. *Nature* **204**, 436-437 (1964).
18. Swallow, J.C. & Crease, J. 'Hot salty water at the bottom of the Red Sea'. *Nature* **205**, 165-166 (1965).

Chapter 8

1. Worthington, L.V. 'An attempt to measure the volume transport of Norwegian Sea overflow water through the Denmark Strait'. *Deep-Sea Res.* **16** (supplement) 421-432 (1969).
2. Swallow, J.C. & Worthington, L.V. 'Deep currents in the Labrador Sea'. *Deep-Sea Res.* **16**, 77-84 (1969).
3. Gould, W.J. & McKee, W.D. 'Vertical structure of semi-diurnal tidal currents in the Bay of Biscay'. *Nature* **244**, 88-91 (1973).
4. Pingree, R.D. 'Flow of surface waters to the west of the British Isles and in the Bay of Biscay'. *Deep-Sea Res. II* **40**, 369-388 (1993).
5. Garrett, C.J.R. & Munk, W. 'Space-time scales of internal waves'. *Geophysical Fluid Dynamics* **80**, 291-297 (1972).
6. Gould W.J. & Sambuco, E. 'The effect of mooring type on measured values of ocean currents'. *Deep-Sea Res. Oceanogr. Abstr.* **22**(1), 55-62 (1975).
7. MEDOC Group. 'Observations of formation of deep water in the Mediterranean Sea, 1969'. *Nature* **227**, 1037-1040 (1969).
8. Wright, P. *Spycatcher: the candid autobiography of a senior intelligence officer* (Viking Penguin Inc., New York & London, 1987).
9. Pollard, R.T., Guymer, T.H. & Taylor, P.K. 'Summary of the JASIN 1978 field experiment'. *Phil. Trans. R. Soc. Lond.* **308**(1503), 221-230 (1983).
10. MODE Group. 'The Mid-Ocean Dynamics Experiment'. *Deep-Sea Res.* **25**, 859-910 (1978).
11. Gould, J. 'From Swallow floats to Argo: the development of neutrally buoyant floats'. *Deep-Sea Res. II* **52**(3-4), 529-543 (2005).
12. Gould, W.J. 'Direct measurement of subsurface currents; a success story' in Deacon, M., Rice, A.L. & Summerhayes, C. *Understanding the oceans: a century of ocean exploration* (UCL Press, London, 2001).
13. Gould, W.J. & Turton, J. 'Argo: sounding the oceans'. *Weather* **61**(1), 17-21 (2006).
14. Cunningham, S.A. *et al.* 'Temporal variability of the Atlantic meridional overturning circulation at 26.5°N'. *Science*, **317**(5840), 935-938 (2007).

Chapter 9

1. Mortimer, C.H. 'Some effects of the earth's rotation on water movement in stratified lakes'. *Proc. Intern. Assoc. Appl. Limnol.* **12**, 66-77 (1955).
2. Thorpe, S.A., Hall, A. & Crofts, I. 'The internal surge in Loch Ness'. *Nature* **237**, 96-98 (1972).
3. Thorpe, S.A. 'Near-resonant forcing in a shallow two-layer fluid: a model for the internal surge in Loch Ness'. *J. Fluid Mech.* **63**, 509-527 (1974).
4. Thorpe, S.A. 'Experiments on the instability of stratified shear flows: immiscible fluids'. *J. Fluid Mech.* **39**, 25-48 (1969).
5. Thorpe, S.A. 'Experiments on the instability of stratified shear flows: miscible fluids'. *J. Fluid Mech.* **46**, 299-319 (1971).
6. Thorpe, S.A. 'A sediment cloud below the Mediterranean outflow'. *Nature* **239**, 326-327 (1972).
7. Thorpe, S.A., Collins, E.P. & Gaunt, D.I. 'An electromagnetic current meter to measure turbulent fluctuations near the ocean floor'. *Deep-Sea Res.* **20**, 933-938 (1973).

Chapter 10

1. Cox R.A & Smith, N.D. 'The specific heat of sea water'. *Proc. R. Soc. London. Series A, Mathematical and Physical Sciences* **252**(1268), 51-62 (1959).
2. Cox, R.A., Culkin, F., Greenhalgh, R. & Riley, J.P. 'Chlorinity, conductivity and density of sea water'. *Nature* **193**, 518-520 (1962).
3. Cox, R. A. 'Temperature compensation in salinometers'. *Deep-Sea Res.* **9**, 504-506 (1962).
4. Cox, R. A., Culkin, F. & Riley, J.P. 'The electrical conductivity/chlorinity relationships in natural sea water'. *Deep-Sea Res.* **14**, 203-220 (1967).
5. Hermann, F. & Culkin, F. 'The preparation and chlorinity calibration of standard seawater'. *Deep-Sea Res.* **25**, 1265-1270 (1978).
6. Rusby, J.S.M. 'Measurements of the refractive index of seawater relative to Copenhagen Standard Seawater'. *Deep-Sea Res.* **14**(4), 427-439 (1967).
7. Cox, R.A., McCartney, M.J. & Culkin, F. 'Pure water for relative density standard'. *Deep-Sea Res.* **15**, 319-325 (1968).
8. Culkin, F. & Smith, N.D. 'Determination of the concentration of potassium chloride solution having the same electrical conductivity at 15 deg. C and infinite frequency as Standard Seawater of salinity 35'. *J. Oceanic Engin.* **5**, 22-23 (1980).
9. UNESCO. 'The Practical Salinity Scale 1978 and the International Equation of State of Seawater 1980'. *UNESCO technical papers in marine science* 36 (1981).
10. *International Oceanographic Tables, Vol. 1* (National Institute of Oceanography of Great Britain & UNESCO, Paris, 1966).
11. *International Oceanographic Tables, Vol. 2* (National Institute of Oceanography of Great Britain & UNESCO, Paris, 1966).
12. Brown, N.L. & Allentoft, B. 'Salinity, conductivity and temperature relation of seawater over the salinity range 0-50'. Final Rep. Contract No ONR4290(00) M.J.O. no.2003 Bissett-Berman, *U.S. Office of Naval Res., Washington, DC.* 1345-1346 (1966).
13. Culkin, F. & Cox, R.A. 'Sodium, potassium, magnesium, calcium and strontium in seawater'. *Deep-Sea Res.* **13**, 789-804 (1966).
14. Brewer, P.G., Riley, J.P. & Culkin, F. 'Chemical composition of the hot salty water at the bottom of the Red Sea'. *Nature* **206**, 1345-1346 (1965).

Chapter 11

1. Cornish, V. *Waves of the Sea and Other Water Waves* (Open Court Pub. Co., Chicago, 1912).
2. Sverdrup, H.U. & Munk, W.H. *Wind, sea and swell: theory of relations for forecasting* (U.S. Hydrographic Office, Publ. H.O. 601, 1947).
3. Ursell, F. 'Wave generation by wind' in Batchelor, G. K. & Davies, R.M. (eds) *Surveys in Mechanics* (Cambridge Univ. Press, Cambridge, 1956).
4. Barber, N.F. 'Measurement of sea conditions by the motions of a floating buoy' in Barber, N.F., Tucker, M.J. & Longuet-Higgins, M.S. *Four theoretical notes on the estimation of sea conditions* (Admiralty Res. Lab. Report 103.40/N2/W., 1946).
5. Longuet-Higgins, M.S., Cartwright, D.E., & Smith, N.D. 'Observations of the directional spectrum of sea waves using the motions of a floating buoy' in Vetter, R.C. (ed.) *Ocean Wave Spectra* (Englewood/Prentice-Hall, New Jersey, 1963).
6. Phillips, O.M. 'The equilibrium range in the spectrum of wind-generated waves'. *J. Fluid Mech.* **4**, 426-431 (1958).
7. Phillips, O.M. 'Spectral and statistical properties of the equilibrium range in wind-generated gravity waves'. *J. Fluid Mech.* **156**, 505-531 (1985).
8. Miles, J.W. 'On the generation of surface waves by shear flows'. *J. Fluid Mech.* **3**, 185-204 (1957).
9. Phillips, O.M. 'On the generation of waves by a turbulent wind'. *J. Fluid Mech.* **2**, 417-445 (1957).
10. Phillips, O.M. 'On the dynamics of unsteady gravity waves of finite amplitude, Part 1. The elementary interactions'. *J. Fluid Mech.* **9**, 193-217 (1960).
11. Hasselmann, K. 'On the non-linear energy transfer in a gravity-wave spectrum. Part 1: General theory'. *J. Fluid Mech.* **12**, 481-500 (1962).
12. Hasselmann, K. 'On the non-linear energy transfer in a gravity-wave spectrum, Part 2. Conservation theorems: wave-particle analogy: irreversibility'. *J. Fluid Mech.* **15**, 273-281 (1963).
13. Hasselmann, K. 'On the non-linear energy transfer in a gravity-wave spectrum, Part 3. Evaluation of the energy flux and swell-sea interaction for a Neumann spectrum'. *J. Fluid Mech.* **15**, 385-398 (1963).
14. Longuet-Higgins, M.S. & Smith, N.D. 'An experiment on third-order resonant wave interactions'. *J. Fluid Mech.* **25**, 417-435 (1966).
15. McGoldrick, L.F., Phillips, O.M., Huang, N.E., & Hodgson, T.H. 'Measurements of third-order resonant wave interactions'. *J. Fluid. Mech.* **25**, 437-456 (1966).
16. Longuet-Higgins, M.S., & Stewart, R.W. 'Radiation stresses in water waves; a physical discussion, with applications'. *Deep-Sea Res.* **11**, 529-562 (1964).
17. Longuet-Higgins, M.S. 'Wave set-up, percolation and undertow in the surf zone'. *Proc. R. Soc. Lond.* A **390**, 283-291 (1983).
18. Bowen, A.J. 'The generation of longshore currents on a plane beach'. *J. Mar. Res.* **27**, 206-215 (1969).
19. Longuet-Higgins, M.S. 'The generation of capillary waves by steep gravity waves'. *J. Fluid Mech.* **16**, 138-159 (1963).
20. Rattray, M. 'Time dependent motion in an ocean: a unified, two-layer beta-plane approximation'. *Stud. Oceanogr, Geophys. Inst. Tokyo Univ.* 19-29. (1964).
21. Longuet-Higgins, M.S. & Pond, G.S. 'The free oscillations of fluid on a hemisphere bounded by meridians of longitude'. *Phil. Trans. R. Soc. Lond.* A **266**, 193-223 (1968).
22. Longuet-Higgins, M.S. 'On group velocity and energy flux in planetary wave motions'. *Deep-Sea Res.* **11**, 35-42 (1964).

23. Lighthill, M.J. 'Dynamic response of the Indian Ocean to onset of the southwest monsoon'. *Phil. Trans. R. Soc. Lond.* A **265**, 46-92 (1969).
24. Longuet-Higgins, M.S. 'Planetary waves on a rotating sphere'. *Proc. R. Soc. Lond.* A **279**, 446-473 (1964).
25. Longuet-Higgins, M.S. 'Planetary waves on a rotating sphere. II'. *Proc. R. Soc. Lond.* A **284**, 40-68 (1965).
26. Longuet-Higgins, M.S. 'Planetary waves on a hemisphere bounded by meridians of longitude'. *Phil. Trans. R. Soc. Lond.* A **260**, 317-350 (1966).
27. Longuet-Higgins, M.S. 'The eigenfunctions of Laplace's tidal equations over a sphere'. *Phil. Trans. R. Soc. Lond.* A **262**, 511-607 (1968).
28. Longuet-Higgins, M.S. 'On the trapping of wave energy round islands'. *J. Fluid Mech.* **29**, 781-821 (1967).
29. Cartwright, D.E. *Tides, a Scientific History* (Cambridge Univ. Press, Cambridge, 1999).
30. Cartwright, D.E., Edden, A.C., Spencer, R. & Vassie, J.M. 'The tides of the northeast Atlantic Ocean'. *Phil. Trans. R. Soc. Lond.* A **298**, 87-139 (1980).
31. Longuet-Higgins, M.S. 'The statistical analysis of a random moving surface'. *Phil. Trans. R. Soc.* A **249**, 321-387 (1957).
32. Longuet-Higgins, M.S. 'The statistical geometry of random surfaces' in Lin, C.C. (ed.) *Hydrodynamic Instability* (Amer. Math. Soc., Providence, 1960).
33. Longuet-Higgins, M.S. 'The effect of nonlinearities on statistical distributions in the theory of sea waves'. *J. Fluid Mech.* **17**, 459-480 (1963).
34. Longuet-Higgins, M.S. 'Breaking waves – in deep and shallow water' in *Proc. 10th Symp. Naval Hydrodynamics, Cambridge, Mass.* 597-605 (1974).
35. Longuet-Higgins, M.S. Donelan, M.A. & Turner, J.S. 'Periodicity in whitecaps'. *Nature* **239**, 449-451 (1972).
36. Longuet-Higgins, M.S. & Turner, J.S. 'An "entraining plume" model of a spilling breaker'. *J. Fluid Mech.* **63**, 1-20 (1974).

Chapter 12

1. Longuet Higgins, M.S., Cartwright, D.E. & Smith, N.D. 'Observations of the directional spectrum of sea waves using the motions of a floating buoy' in *Ocean Wave Spectra*, pp. 111-136 (Prentice-Hall, New Jersey, 1963).
2. Cartwright, D.E. & Smith, N.D. 'Buoy techniques for obtaining directional wave spectra' in *Buoy Technology*, pp. 112-122 (Marine Technology Society, Washington, DC, 1964).
3. Mitsuyasu, H. *et al.* 'Some measurements of the directional wave spectrum'. *J. Physical Oceanogr.* **5**, 750-762 (1975).
4. Hasselmann, K. *et al.* 'Measurement of wind-wave growth and swell decay during the Joint North Sea Wave Project (JONSWAP)'. *Deutsches Hydrographisches Zeitschrift*, A(8), 12, Hamburg (1973).
5. Cartwright, D.E. & Rydill, L.J. 'The rolling and pitching of a ship at sea (with written discussion)'. *Trans.Inst. Naval Architects* **99**, 100-135 (1957).
6. Cartwright. D.E. 'On the vertical motions of a ship in sea waves' in *Proc. Symp. on Behaviour of ships in a seaway*, pp.1-27 (Wageningen, 1959).
7. Canham, H.J.S., Cartwright, D.E., Goodrich, G.J. & Hogben, N. 'Seakeeping trials on OWS Weather Reporter (with written discussion)' in *Trans. R. Inst. Naval Architects*, 447-492 (1962).
8. Ishiguro, S. 'A method of analysis for long-wave phenomena in the ocean, using electronic network models. 1. The earth's rotation ignored'. *Phil. Trans. A.* **251**, 303-340 (1959).

9. Ishiguro, S. 'Electric Analogues in Oceanography' in Barnes. H. (ed.) *Oceanography and Marine Biology: an annual review, Vol. 10*, pp. 27-96 (Allen & Unwin, London, 1972).
10. Cartwright, D.E. & Crease, J. 'A comparison of the geodetic reference levels of England and France by means of the sea surface'. *Proc. R. Soc. Lond. A.* **273**, 538-580 (1963).
11. Munk, W.H. & Cartwright, D.E. 'Tidal spectroscopy and prediction'. *Phil. Trans. R. Soc. Lond. A.* **259**, 533-581 (1966).
12. Cartwright, D.E. 'A unified analysis of tides and surges round north and east Britain'. *Phil. Trans. R. Soc. Lond. A.* **283**, 1-55 (1968).
13. Collar, P.G. & Cartwright, D.E. 'Open sea tidal measurements near the edge of the northwest European continental shelf'. *Deep-Sea Res.* **19**, 673-689 (1972).
14. Report of SCOR Working Group 27. 'An intercomparison of open sea tidal pressure recorders'. *UNESCO Technical papers in Marine Science* **21** (UNESCO, Paris, 1975).
15. Cartwright, D.E. 'Satellite Altimetry in Geodesy and Oceanography' in Rummel. R. & Sanso, F. (eds) *Lecture notes in Earth Science*, pp. 100-141 (Springer, Berlin, 1993).

Chapter 13

1. Draper, L. 'The history of wave research at Wormley: a personal view'. *Ocean Challenge* **6**, 24-27 (1996).
2. Darbyshire, J. 'The generation of waves by wind'. *Proc. R. Soc. Lond. A* **215**, 299-328 (1952).
3. Darbyshire, J. 'An investigation of storm waves in the North Atlantic Ocean'. *Proc. R. Soc. A* **230**, 560-569 (1955).
4. Darbyshire, M. & Draper, L. 'Forecasting wind-generated sea waves'. *Engineering* **195**, 482-484 (1963).
5. Darbyshire, J. 'Sea conditions at Tema Harbour'. *Dock Harb. Author.* **38**, 277-278 (1957).
6. Longuet-Higgins, M.S. 'On the statistical distribution of the heights of sea waves'. *J. Marine Res.* **11**, 245-266 (1952).
7. Cartwright, D.E. 'On estimating the mean energy of sea waves from the highest wave in a record'. *Proc. R. Soc. A* **247**, 22-48 (1958).
8. Tucker, M.J. 'Analysis of records of sea waves'. *Proc. Instn. Civil Engrs.* **26**, 305-316 (1963).
9. Draper, L. 'Derivation of a "design wave" from instrumental records of sea waves'. *Proc. Instn. Civ. Engrs* **26**, 291-304 (1963).
10. Carter, D.J.T. and Draper, L. 'Has the NE Atlantic become rougher?' *Nature* **332**, p.496 (1988).
11. Bouws, E., Günter, H., Rosenthal, W. & Vincent, C.L. 'Similarity of the wind wave spectrum in finite depth water. Part 1: spectral form'. *J. Geophysical Res.* **90** (C1), 975-986, (1985).

Chapter 14

1. Chesterman, W.D., Clynick, P.R. & Stride, A.H. 'An acoustic aid to sea bed survey'. *Acustica* 8, 285-290 (1958).
2. Donovan, D.T. & Stride, A.H. 'An acoustic survey of the sea floor south of Dorset and its geological interpretation'. *Phil. Trans. R. Soc. B* 244, 299-330 (1961).

3. Smith, A.J., Stride, A.H. & Whittard, W.F. 'The geology of the Western Approaches of the English Channel IV. A recently discovered Variscan granite west-north-west of the Scilly Isles'. *Colston Papers* 17, 287-301 (1965).
4. Stride, A.H. 'Current-swept sea floors near the southern half of Great Britain'. *Quart. J. Geol. Soc. Lond.* 119, 175-199 (1963).
5. Belderson, R.H. & Stride, A.H. 'Tidal current fashioning of a basal bed'. *Marine Geol.* 4, 237-257 (1966).
6. Kenyon, N.H. 'Sand ribbons of European tidal seas'. *Marine Geol.* 9, 25-39 (1970).
7. Kenyon, N.H. & Stride, A.H. 'The tide-swept continental shelf sediments between the Shetland Isles and France'. *Sedimentology* 14, 159-173 (1970).
8. Pingree, R.D. & Griffiths, D.K. 'Sand transport paths around the British Isles resulting from M2 and M4 tidal interactions'. *J. Mar. Biol. Assoc. UK* 59, 497-513 (1979).
9. Belderson, R.H., Kenyon, N.H. & Wilson, J.B. 'Iceberg plough marks in the northeast Atlantic'. *Palaeogeogr., Palaeoclim. Palaeoecol.* 13, 215-224 (1973).
10. Stride, A.H. 'Sediment transport by the North Sea' in D. Goldberg, D. (ed.) *North Sea Science*, pp.101-130 (MIT Press, 1973).
11. Stride, A.H., Curray, J.R., Moore, D.G. & Belderson, R.H. 'Marine geology of the continental margin of Europe'. *Phil. Trans. R. Soc. Lond. A* 264, 31-75 (1969).
12. Kenyon, N.H. & Belderson, R.H. 'Bedforms of the Mediterranean undercurrent observed with side-scan sonar'. *Sedimentary Geol.* 9, 77-99 (1973).
13. Belderson, R.H., Kenyon, N.H., Stride, A.H. & Stubbs, A.R. *Sonographs of the Sea Floor* (Elsevier, Amsterdam, 1972).

Chapter 15

1. Heezen, B.C., Tharp, M. & Ewing, M. 'The Floors of the Ocean. 1. The North Atlantic'. *Geol. Soc. Am., Special Paper* 65 (1959).
2. Laughton, A.S. 'A new deep-sea camera'. *Deep-Sea Res.* 4, 120-125 (1957).
3. Hersey, J.B. (ed.) *Deep-Sea Photography* (Johns Hopkins Press, Baltimore, 1967).
4. Heezen, B.C. & Laughton, A.S. 'Abyssal plains' in Hill, M.N. (ed.) *The Sea: ideas and observations on progress in the study of the sea, Vol. 3*, pp.312-364 (Wiley Interscience, London, 1963).
5. Laughton, A.S. 'Microtopography' in Hill, M.N. (ed.) *The Sea: ideas and observation on progress in the study of the sea, Vol. 3*, pp.437-472 (Wiley Interscience, London,1963).
6 Belderson R.H. & Laughton, A.S. 'Correlation of some Atlantic turbidites'. *Sedimentology* 7, 103-116 (1966).
7. Laughton, A.S. 'An interplain deep-sea channel system'. *Deep-Sea Res.* 7, 75-88 (1960).
8. Laughton, A.S., Roberts, D.G. & Graves R. 'Bathymetry of the north-east Atlantic: Mid-Atlantic Ridge to south-west Europe'. *Deep-Sea Res.* 22, 791-810 & Admiralty Chart C6568 (1975).
9. *The History of GEBCO 1903-2003: The 100 year story of the General Bathymetric Chart of the Oceans* (GITC bv, Lemmer, Netherlands, 2003).
10. Laughton, A.S., Roberts, D.G. & Graves R. 'Deep ocean floor mapping for scientific purposes and the application of automatic cartography'. *Intern. Hydrographic Rev.* 50(1), 125-148 (1973).
11. Hill, M.N. (ed.) *The sea: ideas and observations on progress in the study of the sea. Vol 3: The Earth beneath the sea* (Wiley Interscience, London,1963).
12. Maxwell, A,E. (ed.) *The Sea: ideas and observations on progress in the study*

of the seas. Vol 4: New concepts of sea floor evolution (Wiley-Interscience, New York, 1970).
13. Laughton, A.S. 'The Gulf of Aden'. *Phil.Trans.R. Soc, A* 259, 150-171 (1966).
14. Laughton, A.S., Whitmarsh, R.B. & Jones, M.T. 'The evolution of the Gulf of Aden'. *Phil. Trans. R. Soc. A* 267, 227-266 (1970).
15. *Geological-Geophysical Atlas of the Indian Ocean* (Acad. Sci. USSR, Moscow, 1975).
16. Laughton, A.S. *et al.* 'A continuous east-west fault on the Azores-Gibraltar Ridge'. *Nature*, 237, 217-220 (1972).
17. Matthews, D.H. *et al.* 'Crustal structure and origin of Peake and Freen Deeps, N.E. Atlantic'. *Geophys. J. R. Astron. Soc.* 18, 517-542 (1969).
18. Laughton, A.S. & Rusby, J.S.M. 'Long-range sonar and photographic studies of the median valley in the FAMOUS area of the Mid-Atlantic Ridge near 37N'. *Deep-Sea Res.* 22, 279-298 (1975).
19. Whitmarsh, R.B. & Laughton, A.S. 'A long-range sonar study of the Mid-Atlantic Ridge near 37°N (FAMOUS area) and its tectonic implications'. *Deep-Sea Res.* **23**, 1005-1023 (1976).
20. Laughton, A.S. 'The first decade of GLORIA'. *J. Geophys. Res.* **86**, 11511-11534 (1981).
21. Laughton, A.S. 'South Labrador Sea and the evolution of the North Atlantic'. *Nature* **232**, 612-617 (1971).
22. Laughton, A.S. *et al. Initial Reports of the Deep Sea Drilling Project, Vol. 12* (U.S. Government Printing Office, Washington D.C.,1972).
23. Whitmarsh, R.B., Hamilton, N. & Kidd, R.B. 'Paleomagnetic results for the Indian and Arabian plates from Arabian Sea cores' in *Initial Reports of the Deep Sea Drilling Project* **23** by Whitmarsh, R.B. *et al.*, pp. 521-525 (U.S. Government Printing Office, Washington D.C., 1974).
24. Roberts, D.G., Bishop, D.G., Laughton, A.S., Ziolkowski, A.M. & Scrutton, R.A. 'New sedimentary basin on Rockall Plateau'. *Nature* **225**, 170-172 (1970).
25. Roberts, D.G. 'Structural development of the British Isles continental margin and the Rockall Plateau' in Burk, C.& Drake, C.L. (eds) *Geology of continental margins* (Springer-Verlag, New York,1974).
26. Roberts, D.G. 'Marine geology of the Rockall Plateau and Trough'. *Phil. Trans. R. Soc. Lond. A* **278**, 447-509 (1975).
27. Whitmarsh, R.B. 'An ocean bottom pop-up seismic recorder'. *Marine Geophys. Res.* **1**, 91-98 (1970).
28. Whitmarsh R.B. 'Median valley refraction line, Mid-Atlantic Ridge at 37°N'. *Nature* **246**, 297-299 (1973).
29. Whitmarsh, R.B. 'Seismic anisotropy of the uppermost mantle absent beneath the east flank of the Reykjanes Ridge'. *Bull. Seismol. Soc. Am.* **61**, 1351-1368 (1971).
30. Flemming, N.C. 'Archaeological evidence for eustatic change of sea level and earth movements in the Western Mediterranean during the last 2,000 years'. *Geol. Soc. Am. Special Paper* **109** (1969).
31. Flemming, N.C. 'Relative chronology of submerged Pleistocene marine erosion features in the western Mediterranean'. *J. Geol.* **80**. 633-662 (1972).

Chapter 16

1. Tucker, M.J. 'A shipborne wave recorder'. *Trans. R. Inst. Nav. Architects* **98**, 236-250 (1956).
2. Pierson, W.J. & Moskowitz, L.A. 'Proposed spectral form for a fully developed wind sea based on the similarity theory of S.A. Kitaigorodskii'. *J.*

Geophys. Res. **69**, 5181-5203 (1964).
3. Draper, L. 'The history of wave research at Wormley: a personal view'. *Ocean Challenge* **6**, 24-27 (1996).
4. Holliday, N.P. *et al.* 'Were extreme waves in the Rockall Trough the largest ever recorded?' *Geophys. Res. Lett.* **33**, L05613/doi: 10.1029/2005GL025238 (2006).
5. Canham, H.J.S., Cartwright, D.E., Goodrich, G.J., & Hogben, N. 'Seakeeping trials on OWS "Weather Reporter"'. *Trans. R.I.N.A.* **104**, 447-492 (1962)
6. Cartwright, D.E. & Smith, N.D. 'Buoy Techniques for obtaining directional wave spectra' in *Trans. 1964 Buoy Technol. Symp., Marine Tech. Soc.,* Washington D.C. 112-121 (1964).
7. Mitsuyasu, H. *et al.* 'Observations of the directional spectrum of ocean waves using a cloverleaf buoy'. *J. Phys. Oceanogr.* **5**, 750-760 (1975).
8. Lee, A.J., Collar. P.G. & Carson, R.M. 'The U.K. large data buoy programme'. *Proc. Soc. Underwater Technol.* **2**(3), 14-27 (1973).
9. Rusby, J.S.M., Kelly, R.F., Wall, J., Hunter, C.A. & Butcher, J. 'The construction and offshore testing of the UK Data Buoy (DB1 project)'. *Proc. Oceanol. Intern.* **78** Tech. Session J 'Instruments and communications' 64 (1978).
10. Swallow, J.C., McCartney, B.S. & Millard, N.W. 'The Minimode float tracking system'. *Deep-Sea Res.* **21**(7), 573-595 (1974).
11. Hinde, B.J. & Gaunt, D.I. 'Some new techniques for recording and analysing microseisms'. *Proc. R. Soc. A* **290**, 297-317 (1966).
12. Hinde, B.J. & Gaunt, D.I. 'Microseisms'. *Contemp. Phys.* **8**(3), 267-283 (1967).
13. Tucker, M.J., Smith, N.D., Pierce, F.E. & Collins, E.P. 'A two-component ship's log'. *J. Inst. Nav.* 23, 302-316 (1970).
14. Hamon, B.V. 'A portable temperature-chlorinity bridge for estuarine investigations and sea water analysis'. *J. Sci. Instrumen.* **33**, 329-332 (1956).
15. Bowers, R. & Bishop, D.G. 'A towed thermistor chain for temperature measurement at various depths'. Paper No. 7 in *Proceedings of the I.E.R.E. Conference on Electronic Engineering in Oceanography, Southampton, 12-15 September 1966* (Inst. Electronic and Radio Engineers, London, 1966).
16. Tucker, M.J., & Stubbs, A.R. 'Narrow beam echo-ranger for fishery and geological investigations'. *Brit. J. Appl. Phys.* **12**, 103-110 (1961).
17. Belderson, R.B., Kenyon, N.H., Stride, A.H., & Stubbs, A.R. *Sonographs of the sea floor: a picture atlas* (Elsevier, Amsterdam, 1972).
18. McCartney, B.S., Stubbs, A.R. & Tucker, M.J. 'Low frequency target strengths of pilchard shoals and the hypothesis of swimbladder resonance'. *Nature* **207**, 39-40 (1965).
19. McCartney, B.S. & Stubbs, A.R. 'Measurements of the acoustic target strengths of fish in dorsal aspect, including swimbladder resonance'. *J. Sound Vibr.* **15**(3), 397-420 (1971).
20. McCartney, B.S. 'Comparison of the acoustic and biological sampling of the sonic scattering layers: R.R.S. "Discovery" SOND cruise, 1965'. *J. Mar. Biol. Assoc. U.K.* **56**(1), 161-178 (1976).
21. Bary, B.McK. 'The effect of electric fields on marine fishes'. *Scottish Home Department, Marine Research Report* No.1 (1956).
22. Carruthers, J.N. 'How fishermen can measure currents in their own interests and how measurements can easily be made of some features of trawl shape and of warp slope and direction'. *Scot. Fisheries Bull.* **6**, 2-5. (1957).
23. Carruthers, J.N. 'The "Pisa" leaning-tube bottom current indicator: jelly/oil bottle version for free-fall use'. *Bull. de l'Inst. Oceanogr, Monaco* **70** (1418), 15pp. (1972).

Chapter 18

1. Deacon, G.E.R. 'James Norman Carruthers: obituary'. *Polar Rec.* **16**, 873-875. (1973).
2. Gould, J. 'Obituary: Mary Swallow, 1917-2006'. *Deep-Sea Res. I* **53**, 749-750 (2006).

Index

All references in bold refer to illustrations within the text.

Subjects

People

Breinigsville, PA USA
14 September 2010

245373BV00004B/1/P

9 780718 892302